Items should be returned on or before the last date shown below. Items may be renewed by personal application, by writing or by telephone. To renew give the date due and the number on the barcode label. Fines are charged on overdue items and will include postage incurred in recovery. Damage to, or loss of items will be charged to the borrower.

Date Due	Date Due	Date Due
04. JAN 97		
07. FEB 97		
	28. FEB 00	
18. FEB 97		
03. APR 97		
26. APR 97	13. NOV 01.	
	01 DEC 01	
17. MAY 97	01. MAY 02	
02. FEB 98	23. AUG 03.	
	08.	
08.		
23. SEP 99	12 JUL '13	

STONES AND STARS

DENISE HALL

STONES
AND
STARS

A Year in West Cork

M
MACMILLAN
LONDON

First published 1994 by Macmillan

a division of Pan Macmillan Publishers Limited
Cavaye Place London SW10 9PG
and Basingstoke

Associated companies throughout the world

ISBN 0 333 61124 1

The right of Denise Hall to be identified as the
author of this work has been asserted by her in accordance
with the Copyright, Designs and Patents Act 1988.

1 3 5 7 9 8 6 4 2

A CIP catalogue record for this book is available from
the British Library

Typeset by CentraCet Limited, Cambridge
Printed and bound in Great Britain by
Mackays of Chatham PLC, Chatham, Kent

To Amber, for being

'To you is left, (unspeakably confused)
your life, gigantic, ripening, full of fears,
so that it now hemmed in, now grasping all,
is changed in you by turns to stones and stars.'

'Evening' by Rainer Maria Rilke

CONTENTS

CONTENTS

ACKNOWLEDGEMENTS

Special thanks to RTE Radio, the *Cork Examiner*, the *Irish Post*, Deirdre Purcell, Gesina Pelka-Bastion, Jane Wood, Peta Nightingale and the people of West Cork, past and present.

CHAPTER ONE

HOOKED!

Barley Lake Cottage was the last house on a narrow and winding mountain road and its views were magnificent. I had been living there happily for about three years.

It was basically one long room with a huge stone fireplace, three small windows and a single door. Upstairs, the previous owner had skilfully wangled two bedrooms out of attic space. The cottage was about eighty years old and had thick, uneven walls and chunky grey roof slates said to have been brought over the mountains from Kerry by donkey and cart some sixty years ago.

There was no electricity or running water when I moved in, just oil lamps and a well at the bottom of the steep hill outside my door. My back garden was a half-acre field which sloped steeply away from the house, and overlooked the heavily wooded glen. It seemed like a huge amount of space for a garden to me. But then, I had moved to West Cork after ten years' hard time in Los Angeles, where my modest Irish half-acre would have been considered enough land on which to build a mini-mall, a duplex or two and a parking lot.

I

The last three years of my LA stint had been spent working for the notorious *National Enquirer*, and I'd decided it was high time for a change. In this remote spot amongst the peace and tranquillity of the mountains, I hoped that my chaotic life would come to absorb the serenity around me. I also had the notion that somehow I would be reclaiming a part of myself.

Ireland was where my father's people came from, and, although I'd never lived there, over the years I visited often. After only a few days of walking through Killarney's woods and drinking pints in small dark pubs I always felt strangely at home in a way I had never experienced, or even looked for, in other countries where I had lived. Ireland just felt comfortable to me, like slipping into a pair of well-worn slippers whose contours seem to have been designed especially for you.

I had a home here, friends, beautiful surroundings, and the clear mountain air had all but banished the memories of LA's nine million souls and depressing brownish-yellow smogs.

Barley Lake Cottage had undergone several improvements over the years – electricity, a proper water supply and, wonder of wonders, a flushing toilet. Admittedly, the toilet was outside, but it struck me as the height of luxury after the wood-chips and Portaloo system I had lived with for several months. Eventually I planned to add a bathroom and separate kitchen to the little cottage. Then it would be perfect. But those luxuries would have to wait.

I was struggling to earn a living as a freelance journalist based, as many editors seemed to think, in the furthest reaches of nowhere. In fact, some of them appeared to find it hard to believe that anything of interest to the outside world ever happened here. I found this attitude hard to understand. Personally, I thought the pace of country living was relentless. There always seemed to be so much going on and I had already realized if I wanted a break from this frenetic lifestyle

I had to go to a nice quiet city to get it. And that was before
I bought Lickeen.

It began innocently enough – a chance encounter with
the local auctioneer in the forest. He was taking a constitu-
tional, I was escaping from a feature I should have been
writing, on the pretext of walking the dogs and feeding my
two horses.

The auctioneer and I chatted pleasantly for a while, the
weather-and-sheep sort of conversation that forms a comfort-
ing backdrop to so much of daily life here. I happened to
mention that my daughter Amber's oldest and closest friend
Kristine, who lives in Los Angeles and visits us often, had
become so enamoured of the pristine beauty of West Cork
that she was thinking of buying a small piece of land in our
glen, situated a few miles from the coastal village of Glengarriff. Did he know of anything suitable? I wondered.

A familiar gleam appeared in his eye. I should have been
warned. It was a look I had seen before. It meant that he
knew he was on to something, even if I did not.

'What you should do', he said after a moment's thoughtful silence, 'is buy Lickeen.'

I explained again that our friend was looking for something small and uncomplicated and that I certainly had no
intention of moving. I knew all about Lickeen. Everyone in
the glen did. It was seventy acres of beautiful but wild and
neglected land that hadn't been a working farm for thirty
years or more. It had no place in my vision of a well-ordered
life far from the distractions of the city, devoted exclusively
to the serious business of writing. The farmhouse at Lickeen
was derelict, perched halfway up a mountain, the track to it
undriveable. There was no water or electricity and it was
obviously going to take massive amounts of time and money
to make the place habitable. It wasn't the kind of project any
sane person would take on alone.

The auctioneer waited for me to finish. Obviously he
intended to give no quarter.

'What you should do', he repeated with the same, slow thoughtfulness, 'is buy Lickeen.'

I decided to abandon the conversation temporarily. Obviously his mind wasn't on small, insignificant pieces of land right now. We said our goodbyes and I made my way home, the dogs bounding joyfully ahead of me through the rapidly darkening forest, sure I'd heard the last of Lickeen.

Daylight faded fast at this time of year. It would be Christmas soon and my daughter Amber, who was studying psychology at Plymouth University, was coming home for the holidays with her boyfriend Bryn. Kristine would be flying over from Los Angeles as well. Over the last few years, the four of us had become a sort of informal family unit. Although I hadn't managed to find a suitable piece of land for Kristine, something was bound to turn up eventually. There was no great rush.

One of the best things about Christmas in Ireland is that there is absolutely no point in worrying about mundane aspects of life such as bills and the cost of animal feed. For several weeks everything is either closed down or people simply aren't in the mood for serious discussion. From the beginning of December I had discovered you are able to postpone everything by using the magic phrase 'after the holidays'. So now I could legitimately put off thinking about how I was going to increase my meagre income until the New Year. Meanwhile, Lickeen remained unsold and uninhabited except for the badgers, foxes and other assorted wildlife who had claimed it for their own.

The day after the 'family' arrived, we were all sitting round the kitchen table in cosy Barley Lake Cottage, drinking coffee and catching up on everyone's news. It was Christmas Eve, a clear and lovely day, and surprisingly warm. Glengarriff is situated on the Gulf Stream and the weather in early winter is often glorious. Over the years, Amber and I had fallen into the habit of going for long and preferably wild walks on Christmas Eve. It is something we plan with a seriousness and sense of purpose we rarely accomplish at any other time. This year, we decided we would all walk to Lickeen. It had the right elements for a grand Christmas Eve ramble. If we were up to it, we could clamber up the mountain and look down on Bantry Bay with its dark green heavily wooded islands strung out like small perfect jewels far beneath us.

So we did, then we lay exhausted on an outcrop of rock which jutted over some of the tallest and oldest fir trees in Glengarriff Forest, breathing in the heady scents of pine and heather, silently absorbing the surrounding splendour.

The air was so still that day that I could hear the steady effortless beat of a raven's wings as it circled lazily above us. Then someone said that this was a perfect Christmas Eve and Lickeen one of the most beautiful spots on earth. I don't really remember much after that. The next thing I knew, we were all discussing its purchase as calmly as if we were trying to decide between salmon and goose for Christmas dinner.

To this day, my daughter insists that it was me who first mentioned buying the place, but I still find that hard to believe. Anyway, once the idea was released into the crisp winter's air, it just refused to go away. Fuelled by generous amounts of duty-free and the bonhomie of the festive season, we sat planning our future at Lickeen over long, leisurely meals. I would sell Barley Lake Cottage, Kristine would contribute some money, and Amber and Bryn would buy in when they had both finished their degrees. We would start a deer-breeding programme, run orienteering courses, arrange sky-diving lessons off the cliffs. Exactly how we planned to do any of this, with little money or expertise, wasn't quite so clear.

A few days after Christmas, we went down to the auctioneer's office in the village, in a state of great excitement. He had, of course, been expecting us. Our rash decision to buy Lickeen came as no surprise to him.

We agreed that Barley Lake Cottage would go on the

market at Easter. Meanwhile, since I was the only one living in the glen, I would also be responsible for the legalities involved in buying Lickeen. We shook hands with the auctioneer and then went for a couple of drinks to celebrate. I was glad of the opportunity to sit down. I had begun to feel slightly light-headed. I remembered having read recently that moving house was one of the most stressful experiences a person could undergo. Lickeen, with its rambling, wild acres of land, whose boundaries would have to be verified, would not be bought quickly and there would be massive amounts of paperwork involved. But it was too late to turn back now. I sipped my hot whiskey gratefully.

Then suddenly the holidays were over. The family departed with fervent promises to return in the summer when, they said, they would hack rendering and cut back the creeping rhododendrons which some misguided nineteenth-century traveller had introduced here. Now they had naturalized and were in the process of taking over vast areas of land. They would also, they assured me, begin transplanting the twenty-seven acres of Sitka spruce forest we now owned, all of it, as far as we were concerned, planted in the wrong place.

I didn't particularly enjoy being reminded just then of the massive 'To Do' list. It brought me down to earth with a frightening jolt. What had happened to my plans for a period of solitude and writerly contemplation now that I was living alone? Instead of simplifying my life here, I was complicating it in a spectacular and improbable fashion. The others would soon be jetting off to exotic parts, and running water, while I would be left here, wondering where on earth to start.

After they had gone I sat in my silent cottage, contemplating the wilting Christmas tree and a bank balance already teetering dangerously towards the red, wondering for one shell-shocked moment if there was any chance that the *National Enquirer* would consider giving me back my old job.

Lickeen was beautiful. I wanted to live there very much,

and it was going to solve my grazing problems, I reminded myself firmly, trying for the moment not to think about the numerous complications it was about to create in my life. I had two horses to feed, and the half-acre of grass at Barley Lake never lasted long when they were grazing on it. Thanks to the kindness of neighbours and the odd piece of rented land, we managed, but it was getting increasingly difficult.

My vision of a disciplined and orderly life had begun to fade long before I bought Lickeen. It had begun to disintegrate from the moment I had first looked into the deep, dark eyes of Kitty, my twenty-two-year-old ex-forestry mare.

CHAPTER TWO

KITTY AND CO

I had no intention of buying a horse, let alone one of Kitty's awe-inspiring proportions. But she was special, an Irish Draught/Clydesdale cross of gentle and willing disposition who, for most of her twenty years, had worked tirelessly, hauling timber out of Glengarriff Forest, until one day she became redundant in what was an increasingly modernized industry.

I acquired the habit of visiting Kitty when I was out walking with Tom and Sam, my breeding pair of Golden Retrievers. Until Kitty had come into my life, they seemed like livestock enough. They were affectionate and loyal companions who paid their way by obligingly producing a perfect litter of pups every year.

Kitty's field in the forest was near a popular picnic area and a large natural pool, so she often had a lot of visitors. It was always a delight to see her giant head bent gently over the palm of a small child who was tentatively offering her a tit-bit.

When I heard that she would no longer be used for hauling logs out of the forest – it would be strictly JCBs and tractors from now on – I began to wonder what would happen to her. She was getting on a bit, and she had started to suffer from occasional bouts of lameness. She had never had a foal either, so there was little known value in her for breeding. Rumours were rife concerning Kitty's future – she would go for dog-meat, or be left to die where she stood then simply covered over. It just didn't seem fair somehow. After all, she wasn't an old tractor to be abandoned in a ditch. She was a fine, and to me, heart-stoppingly beautiful animal who had always done her best during her long working life. To see Kitty's huge bulk fly gracefully across the field when she saw me coming with an apple was a rare pleasure, one that couldn't have been matched by watching the sleekest racehorse in full gallop.

Perhaps it was because I knew she was the last in a long and noble line, the working forestry horse, or the warm, sweet smell of her on a cold winter's morning when the dogs and I would visit her – in any event, I realized that I couldn't bear the thought of abandoning her. Something had to be done because she seemed to have been forgotten, spending most of what had been a hard, cold winter in her field which, by then, had little grass left in it. She looked more dejected every time I saw her.

I mentioned her plight to some kindly people I knew who lived on the other side of the village. They had several acres of land and a caravan rental business. Since they were regular visitors to her field, they were horrified to hear that Kitty's future seemed so uncertain and promptly contacted the relevant official in the Department of Forestry with an offer to buy her. Soon Kitty had a fine new home, and, so I heard, was believed to be in foal after a visit to an Irish Draught stallion. Her future seemed secure at last.

Some months later, I had an unexpected phone call. The people who had bought Kitty were returning to Dublin. Their business had not been doing well and they were selling up. They wanted to know if I would take Kitty. And so, resolutely ignoring such practicalities as finances and grazing – you need at least four acres per horse – I became Kitty's proud new owner.

To produce a first foal at twenty is unusual to say the least, but as I was to discover, Kitty was no ordinary horse. A thoroughly modern mare, she had her career first before finally settling down to the serious business of motherhood. It was hard to detect her advancing pregnancy because Kitty is anyway broad of girth and, since she had never foaled before, her muscles were still firm. During those interminably long months, I watched over her anxiously. Every day I brushed her dark chestnut coat until it gleamed, cleaned her large feet out with a hoofpick and spent considerable amounts of time leaning on the fence gazing at her admiringly.

The Department of Forestry had agreed to let me rent her old field, and occasionally, Kitty would spend a few days up at Barley Lake with me. Then my sleep would be permeated by the heavy thud of her feet as she ambled around the outside of the house during the night.

I had formed a deep and abiding attachment to this gentle giant and could not imagine being without her. As her pregnancy progressed, so did my anxiety level. At that point, I was completely inexperienced at birthing mares, especially elderly ones, so I had become an avid reader of books on the Horse and had accumulated a considerable library. Because of Kitty's advanced years, the vet suggested I check her every four hours during her last month and I did, staggering round in a sleep-deprived haze for much of the time. The suspense was awful. Mares' pregnancies usually last for eleven months. But not always. Kitty's was rapidly edging towards the twelve-month mark.

One night I was on watch, reading a horse book as usual, about halfway through the section on birth when the author calmly announced that mares can sometimes expel their wombs during foaling. If this should happen, he advised reasonably, you simply scooped the offending organ up in a handy clean sheet and called the vet who would just pop it back in again. No problem.

Except for a rather large one. This handy hint was proffered by a Newmarket vet I noticed, who had the benefit of centrally heated stables, electric light and a bevy of eager and experienced grooms to help him. I had the cold, dark and probably rainy mountain and a torch that invariably refused to work when it was needed. Finding a clean sheet in such conditions, to say nothing of getting to a phone, could prove tricky.

If Kitty's womb had decided to fall out on that first occasion, I had no doubt that it would not be of the dainty, racehorse variety the author was used to. Commensurate with her great bulk, I was certain it would be the size of the dining-room table. I shut the book firmly and stuffed it under a growing mound of ironing where there was little likelihood of it ever being discovered.

The whole thing went on for so long that I became convinced Kitty was going through some sort of elaborate phantom pregnancy. Mares, I was constantly being reminded, are notoriously unpredictable in these matters. Then one morning I stopped off to feed her on my way to Bantry. I was already running late and Kitty was not waiting at the gate as usual. Feeling my irritation rising, I stomped across the field, swiping at the midges which swarmed eagerly around me. I called her and rattled the feed bucket but there was no response.

Then I slipped and nearly fell flat on my face in the large quantity of afterbirth that Kitty had strategically placed on the path. This could only mean one thing. I scanned the field frantically and finally saw what looked like two horses under a distant tree. One of them was Kitty all right, but the other one couldn't be a foal, surely? It looked too big, about the size of a neighbour's pony that sometimes paid Kitty impromptu visits.

I broke into a run, then suddenly remembered the gravity

of the situation and managed to slow my pace down to a walk. As I got closer, I could see that it really was her foal. He was a tottering, wobbly bright chestnut colt, with Kitty's white blaze and a head the size of a moose. He was still streaked with blood and couldn't have been with us all that long but already he was up on his impossibly long legs and suckling enthusiastically.

'You clever old thing,' I whispered, feeling completely awestruck.

Kitty switched her tail as her foal butted greedily at her bulging udder. She gazed dreamily into the distance. I swallowed hard and grinned idiotically. He was perfect, and so very beautiful. I named him Anlon, which means Great Champion. Anlon was the brother of the famed eleventh century Munster King, Brian Boru, and had been a mighty warrior in his day. It seemed a fitting choice of name for this strapping fellow.

I had always been deeply attracted to that unknown world animals inhabited, where so much seemed to be implicit and did not require words. But before Kitty, my experience of the larger species had been sadly limited to the odd donkey-ride on Blackpool beach or an occasional horse-trek in America's High Sierras. This had done little to prepare me for the robust vitality of a colt foal.

Anlon grew at an alarming rate and his appetite quickly matched that of his mother. His wild high spirits seemed to have reduced Kitty, normally so staid and dignified, to a giddy second childhood. The pair of them would fly round the field together, playing foal-type games of side-kicks and bottom pinching for hours.

When Anlon was only nine days old, he went off with Kitty to visit Silverstone, the stallion who had sired him. All mares experience this 'foal-heat'; it is considered the optimum time for conception by breeders and I badly wanted another foal from Kitty. Anlon didn't enjoy the experience much, but Kitty would be with the stallion for about a week, and he was too young to be left alone. He deeply resented his father's amorous attentions towards the mother he adored and, instead of being held by a lead rope close to his mother while the covering took place, as is usual, Anlon kept trying to head-butt his giant of a sire at all the wrong moments. Eventually he had to be shut away in a separate box.

By the time Anlon was about six months old, he had entered what I came to think of as his lager-lout phase. Suddenly, he was wild, wilful and hard-headed, and delighted in thundering up behind me as I picked my way cautiously across the field with a feed bucket, making me jump out of my skin when, with a delighted toss of his ginger mane, he would race off again.

I'm sure that my inexperience with these coltish tactics added considerably to the piquancy of the experience for him. I'd quickly learned that no matter how hard you try to put on a brave face in situations you do not know how to handle, it is pretty hard to fool a horse. They always seem to know and quickly turn your uncertainty to their own advantage.

I wasn't Anlon's only victim. Tom soon became a favourite target too. The dogs always came with me to feed the horses and, generally, neither took much notice of the other, apart from the occasional brief sniff. Tom is a dog who often appears to have a lot on his mind, weighty matters at that if his solemn expression and rigidly staring posture are anything to go by. Whenever Tom sat musing in a corner of the field while I saw to the horses, Anlon would creep up behind him, terrify the poor dog with a quick head-butt, then herd him into a corner where Tom would remain trapped until I rescued him. Sam, who is of a different and entirely more frivolous nature despite being a mother of thirty-eight, had no such problems. She always kept a wary eye on Anlon and seemed to recognize that this wild phase would eventually pass.

As Anlon grew, he turned into a proud and beautiful

animal. He had been weaned from his mother at six months – a traumatic two-week experience when mother and foal must be separated – but they still enjoyed each other's company enormously and usually grazed together. I sold Anlon to a German family who were besotted with him and were planning on moving permanently into their holiday home near me sometime next year. Meanwhile, I happily continued to care for him. When I walked Anlon on a lead rope, essential handling for future training, it was pure joy as he stepped out proudly beside me. That was, of course, on those days when he was not trying to career off in pursuit of important business of his own, trailing me behind him.

If Kitty was in foal again after her second visit to the stallion, this time I hoped it would be a filly so I could breed from her too. Keeping Kitty's line going was becoming increasingly important to me. There was no doubt about it, this heavy-footed animal had changed my life. I spoke knowledgeably of impoverished grazing, laminitis and colic, and had taken to visiting horse-fairs and fantasizing over future purchases. And I devoured the *Farmer's Journal* every week, particularly the section in which horses for sale were advertised.

Thanks to Kitty, I was well primed by the time Lickeen came into the picture.

CHAPTER THREE
PROUD OWNERSHIP

The process of buying Lickeen with its many acres, some of which seemed to keep appearing and disappearing according to which map you were looking at, proved to be every bit as complicated as I had suspected. It began in early February, with detailed title searches of the Land Registry, but I had already been involved in lengthy negotiations over price with the auctioneer and the man who owned Lickeen. He lived in Wexford, and had inherited the place when its former resident, his uncle, had died at the ripe old age of eighty-six. Two years earlier, not believing that anyone would want to live there again, he had planted twenty-seven acres of Sitka spruce. These hardy conifers pretty much look after themselves, and many small farms no longer considered economically viable have been lost to them.

Eventually we came to an agreement on price, one that took into account the derelict condition of the house and the land, which was poor if you didn't want to spend a lifetime watching the forest growing up around you. And I did not. But it would be two more years before I could get to work on reclaiming the former pastures where most of the Sitkas were because the trees had been planted under a grant incentive scheme, and there was a last payment due to the planters in 1995. At that point, the control of the trees would revert to us. I wasn't too worried about this delay. I knew already that there was going to be more than enough to keep me occupied until then. Meanwhile, the horses would have plenty of rough grazing.

From February to the beginning of April, time seemed to drag interminably as lawyers sorted through documents and surveyors filed reports. There was one five-acre field that was shown on some maps but not on others. I spent several days walking around the place trying to find out if it really existed. It did. I learned from the solicitor that some years ago the farm had been bigger, but small plots of land had gradually been sold off, resulting in some major discrepancies between maps.

The thought of owning this wild and lovely tract of land sometimes woke me in the middle of the night, and I lay filled with a mixture of awe and terror. I experienced moments of intense doubt about my own sanity and I was becoming increasingly concerned about my future as a writer. There was going to be so much work to do on the old farmhouse – knocking off the decayed and crumbling rendering for a start. I couldn't see that there was going to be much time left for being creative. Finances being what they were, I had already decided to do as much of the labouring work as I could manage.

When people asked me, as they frequently did, why I had embarked upon such a foolhardy venture in the first place, I never knew quite how to answer. The truth was that I didn't know. The only thing I was sure of was that I wanted Lickeen more than I had ever wanted anything in my life. There were times when I suspected that it was Lickeen that had chosen me. And, having found me, it wasn't about to let me go.

Apart from keeping on top of the growing mound of

paperwork concerning Lickeen, I had to think about getting Barley Lake Cottage ready for sale. We had agreed on a completion date of 16 June for Lickeen, when all the money would have to be paid, so it was essential that Barley Lake Cottage was sold quickly. It was my only asset.

I painted it inside and out, and tidied the garden. One thing about having horses was that you never had to worry about cutting the grass, or anything else for that matter. Between them, Kitty and Anlon had done an expert job of keeping everything down.

Eventually I was ready. Barley Lake Cottage was bright and welcoming and details of my cosy little home began circulating to potential buyers. People started coming to view it; three different families over one weekend. Then one morning, a friendly young German couple arrived. As soon as they walked through the door, I had a feeling that Barley Lake Cottage had found its new owners.

I showed them round, recognizing only too well the rapturous expressions and glazed look in their eyes. They were hooked. By the next day, they had made me an offer which, after some deliberation, I accepted. It was about five thousand pounds below the price I had been asking but it was still fair, and I knew there was no danger of these two backing out. I asked for a completion date of 16 June in the vain hope that selling Barley Lake Cottage could be co-ordinated with buying Lickeen. But I didn't really expect things to work out quite so neatly. In my experience, they rarely do. I would probably have to get a high-interest bridging loan and go one step deeper into the red with the bank.

It was a strange few weeks, walking round the little house I was so fond of, knowing it wasn't really mine any more. I would miss Barley Lake Cottage but, more than anything now, I wanted to be at Lickeen.

May, as it often is in West Cork, was a glorious month, soft, balmy, and the hedgerows thick with fragrant whitethorn blossom. At night, asleep at Barley Lake Cottage, I dreamed of Lickeen, of how it would be when I lived there, its moods and stillnesses haunting me. As June drew nearer, I was champing at the bit.

But as I had suspected, the sale of Barley Lake Cottage was not going to go through as soon as I had hoped. Some vital piece of paperwork had gone missing and now it looked as though it would be August at the earliest before the transaction was concluded. So I took a deep breath and arranged a bridging loan. I was ready to proceed.

Contracts were solemnly exchanged in the solicitor's office. I was so excited that I spelt my name wrong, to the solicitor's amusement. I didn't even care about the loan any more. Tomorrow, for the first time, I would walk up the boreen knowing that for as far as the eye could see, the land was ours – mine, Amber's, Bryn's and Kristine's. It was a humbling thought, especially since I had always had grave doubts about the concept of ownership. I had come to believe that such things as land and children are only loaned to us to caretake for limited periods of time and that we must do the best we can to live up to the responsibility.

Sleep did not come easily that night. I tossed and turned restlessly. Before it was properly light the next morning, I was up, dressed and ready to go. As dawn broke over the glen, I trudged up the rough track to Lickeen, a backpack containing essential supplies slung over my shoulder, carrying a small hatchet in one hand.

CHAPTER FOUR

HATCHET JOB

I had it on reliable information that a hatchet was the best implement for knocking off the many tons of rendering which had to be removed from my new home. This was to be my first task in the weeks ahead. It had to be done before the bad weather came and it was likely to take quite some time. Because the sale of Barley Lake Cottage hadn't been finalized yet, I still had somewhere civilized to live, which was just as well. I had the feeling that Lickeen was not going to be what you might call comfortable, or even habitable, for quite some time.

The glen was still silent. Patches of mist drifted lazily over the jagged peaks of the distant Caha mountains and the ground was damp and soft from a heavy dew. I was leading Kitty up the track. Her second foal was due in early September and she had come to do her part by grazing down the abundant vegetation. Anlon, to his great disgust, had been left behind in the garden at Barley Lake where the two of them had been spending a couple of days. None of the fields at Lickeen were fenced and he would almost certainly have got himself lost. Kitty was another matter. As long as there was grass, I knew she would stay close to the house.

Tom and Sam raced around us ecstatically in that state of advanced bliss to which they are occasionally prone. Here was a completely fresh territory of dens, setts and burrows, and it was all theirs. I felt a bit like them myself – eager, enthusiastic and with a whole new world to explore.

Then suddenly I came to a halt, and stared around me in disbelief. It was as if I was seeing this vast, neglected tract of land – *really* seeing it – for the first time. My exhilaration evaporated, leaving a sense of helplessness and total inadequacy.

What in God's name was I doing here? One woman and a hatchet against this untamed wilderness? I had clearly bitten off about sixty-nine and a half more acres than I could chew. My heart started to pound alarmingly and I could feel my pulse racing. This was a fully fledged anxiety attack and it was a wonder that it hadn't happened sooner. I wasn't going to be able to cope. I wanted to run away as far and as fast as I could.

Questions I had successfully managed to push to the back of my mind for weeks suddenly demanded instant attention. How were we going to get building supplies up a boreen that was a quarter of a mile long and rough enough to see off an armoured tank? What about the water? When you are mixing cement, you need a lot of it. It would be a long walk from the stream to the house with a full bucket. What was more, I realized that, even with all this land, I still hadn't solved the problem of grazing.

The pastures at Lickeen were criss-crossed by deep drains, dug when the Sitka spruce had been planted. They looked like many highways to nowhere, and were potentially dangerous to young stock like Anlon. Why hadn't I thought of that before? And while I was indulging in a bit of belated self-flagellation I might as well admit that my new home was unliveable in, even by my fairly flexible standards.

I could see nothing ahead of me but a future of back-breaking work, bills, and more back-breaking work. I could have been having a nice, quiet mid-life crisis now, or working in a busy newspaper office and living in a neat one-roomed flat over a hairdresser's instead of being out here, playing Jane of the Mounties. I must be mad.

Kitty looked at me enquiringly for a moment, wondering why we had suddenly stopped, then began munching on a handy clump of fuchsia. The dogs were whiling away the time taking it in turns to roll on the badly decomposing corpse of a badger they had just discovered. At least Lickeen seemed to be living up to someone's expectations, I thought glumly.

A sleek young blackbird perched himself comfortably in a nearby whitethorn bush, plumped out his gleaming feathers then opened his small beak as wide as it would go and belted out an exuberant welcome to the new day. His song positively oozed sweetness and light. So black was my mood that, for a moment, I was tempted to throw a stone at the bush to shut him up. But I thought better of it. He was probably right anyway. There was nothing for it now but to put one foot in front of the other and keep moving.

Kitty's neatly shod feet made a satisfying sound as we crossed the old bridge made of massive, lintel-shaped stones. Beneath us, the stream splashed and gurgled. The boreen divides at this point, skirting a small dark wood full of ancient oaks festooned with moss and ferns. Then it rises steeply up the mountain to where the land is open and bright with tawny-coloured fionnàn and yellow, coconut-scented gorse. The stream is darker there, muttering its way between huge boulders which came to rest millennia ago, when the Ice Age receded and Ireland's giant reindeer still stalked these lands.

Sometimes, up here where no sea ever churned, you can find shells and the fossilized bones of long-dead fish. This is because long ago, when such things were still possible, a giantess with wild red hair that swept the ground lived hereabouts. She was a solitary creature who shunned the company of the puny mortals living below, visiting the nearby sea at night when no one could see her. Her only passion in life was fish, any fish, which she devoured in great quantities as she made her lonely way through these mountains, scattering bones untidily as she went.

The left-hand fork of the boreen rises steeply on its way to the house. Sooner or later, everything round these parts rises steeply. Here the track is dominated by a huge, dark cliff. When it rains heavily, water cascades down its sheer flank and floods the track. When I first came across that primeval slab of stone, I remember smiling a small, involuntary smile of recognition, as you might when visiting an old haunt to discover that a familiar and much-loved object is still there. It was an odd response and one I've never fully understood.

As I passed the cliff on my first morning of ownership, I had the fleeting but powerful sense of *déjà vu*. Countless horses who looked much like Kitty had plodded across the small stone bridge, the men and women who had charge of them walking patiently alongside, thinking perhaps about how to reclaim a difficult field or wondering who would come to help raise a new roof on the barn. There really was nothing new on the face of the earth, I decided, feeling somehow comforted by this revelation.

As I rounded the last bend in the boreen and finally confronted the old house, my rekindled sense of optimism was severely tested. It looked so much worse than I had remembered. Rotting window frames and disintegrating doors clung limply to the crumbling rendering like decaying teeth. Where window frames had given up the unequal struggle and simply fallen out, they had been replaced by chicken-wire nailed haphazardly over the openings. This atmosphere of gloomy decay was further enhanced by the early summer growth of tall dank grasses, weeds and tangled clumps of briar which choked the front of the house. The three majestic sycamores which flanked the house's far side had slid their sinuous limbs halfway across the slates and the chimney-stack was cracked right down the middle, looking in imminent danger of crashing through what was left of the roof.

I tied Kitty to a tree, put my hatchet and backpack down and walked up to the house. I removed some chicken-wire from a window. A sizeable part of the window-frame came with it. The dogs watched me curiously. Then I moved a few rocks from one place to another and hacked at a thick clump of briar with the lethally curved slasher I'd carried up on an earlier visit. I stood back to admire my handiwork and quickly realized that in the face of so much dereliction, it had made absolutely no difference.

The house was approximately a hundred and eighty years old, a two-storey structure with a low-pitched slate roof that I suspected had once been thatched. There were five windows at the front, none at the back, an almost totally collapsed front porch, which led to a still-sturdy door into the kitchen, and a flimsy back door which had rotted so badly that the dogs could fit underneath it. The rendering was discoloured and cracked. Beneath it were tantalizing glimpses of stone. If the stonework was good throughout, I would leave it exposed, as it would have been when the house was first built.

Behind the house, almost completely concealed by tall growths of briars, were the ruined outbuildings – an old stone barn and a tiny one-roomed cottage that had been the first dwelling place at Lickeen. It was estimated to be about four hundred years old, and had an even smaller barn attached to it. None of these buildings had been used for many years. Trees grew up through their earth floors and thick, gnarled ivy clutched possessively at their crumbling stonework. One day this was where my livestock would be housed, I told myself stoutly.

I untied Kitty and let her go. She began munching

enthusiastically. The dogs were asleep, stretched out beneath one of the sycamores, exhausted by the morning's excitement, dreaming, no doubt, of the decomposing badger and the visits they would pay to what was left of it in the future.

It was no good. I couldn't postpone the main event for much longer. I had come to knock off rendering, and God knows, there was enough of it. Grabbing the hatchet, I strode towards the far gable-end wall. It seemed as good a place as any to begin. The wall loomed up before me in Everest-like proportions, but I gave it a good whack with the hatchet anyway.

That first blow echoed reassuringly around the glen. A lump of mortar about the size of a duck-egg fell to the ground at my feet. Well, at least it was a start. I took aim for a second time. Then the midges started to bite.

CHAPTER FIVE

LET THERE BE LIGHT

A hard-pressed traveller to these parts once observed that the West Cork midge must be a cross between insect and piranha fish. When a cloud of the tiny, blood-sucking critters descends on you in full feeding frenzy, madness is not far away. There's something distinctly unsporting about their attack. Silent and practically invisible, they work their way into clothing and hair unnoticed, until you are suddenly swiping at your body with demented fury.

I discovered that the only way to keep this horde at bay was to work at such a furious pace that the huge clouds of dust raised acted like a smokescreen.

By mid-morning I had uncovered an area about the size of your average dining-room table and my right arm felt as heavy as lead. But the stonework I was slowly revealing was spectacular.

It was high time for a break, I decided, putting the hatchet down. I'd make myself a large mug of tea with the camping kettle I'd brought up in my backpack. After I'd checked on Kitty, I'd drink my tea sitting on the cliff which rose sharply on one side of the house, affording sweeping views over the glen. Against this magnificent backdrop, Project Lickeen wouldn't look quite so intimidating.

Kitty didn't come ambling over as she usually did when I whistled and there was no sign of her in the surrounding fields. I decided she would just have to wait until after my break. I knew she wouldn't have gone far. I climbed over a pile of rubble and went through what was left of the back door to put the kettle on.

Kitty seemed to fill the whole kitchen. She was resting comfortably and as she dozed, her rubbery lips flapped steadily in time with her breathing. This was a habit I should promptly discourage. If I didn't, it was going to be difficult convincing Kitty which was her stable and which was mine.

Perhaps she had simply been in search of relief from the maddening insects inside the cool dark kitchen? I had to admit that I rather liked seeing her solid presence there. It made the place look lived-in. At her time of life, and in foal again, she was entitled to an undisturbed nap, I rationalized wildly, as I tend to do with Kitty.

I tiptoed past her slumbering form and turned on the gas cylinder beneath the kettle. I knew from experience that the water would take ages to boil. There'd be plenty of time for a quick wash in the stream and to check the post as well. When I saw the postman's van on the forest road, I remembered a letter I'd been waiting for with a certain amount of dread. He handed over an official looking envelope.

I knew installing electricity was going to be expensive. The house was well off the main road where the power-lines ran. But nothing could have prepared me for the figures which swam before my eyes when I tore open that envelope. Over five thousand pounds . . . There must be some mistake. They had obviously confused me with someone else, who, for reasons best known to themselves, wanted a supply from West Cork to the Australian Outback. Still clutching the envelope, I sat down heavily on the ground, staring bleakly ahead of me into a future of smoking oil lamps and dripping candles.

My retriever Tom suddenly appeared from nowhere and, as he tends to do when he knows I'm upset, placed his large and sympathetic head on my knee. I rubbed his silky ears distractedly. Day one, it wasn't even dinner-time, and disaster had struck. I sat on the grassy verge next to the forest road which runs past Lickeen, leaning against my car and stroking Tom, trying to form some kind of contingency plan. A generator perhaps? But they were expensive, very noisy and I'd always heard stories about them breaking down.

What about my word processor? On the strength of the imminent sale of Barley Lake Cottage I had bought one, trading in my entire collection of decrepit portables. I had felt purposeful, adult even, about the purchase. I would be able to keep copies of stories instead of having to dash thirty miles to the nearest machine at Bantry if anyone requested a duplicate. I would no longer have to feel embarrassed about the way my work looked – never lined up properly with a sickly-looking typeface which produced the sort of text you expected to be signed 'Disgusted, Tunbridge Wells'. But what was the use of a word processor with no electricity? I wouldn't be able to write at all.

Just as I reached this depressing conclusion, a young, cheerful-looking couple drew level with me, tourists on their way to Barley Lake, I guessed. They didn't seem particularly surprised to see me slumped in the ditch muttering to myself. They carried sturdy-looking, indentical backpacks, wore sensible, identical hiking boots and exuded a well-scrubbed healthiness. The dogs welcomed them to our corner of the glen with their customary enthusiasm. (This is a function they take quite seriously, throwing themselves upon all visitors as if they are long-lost friends from whom it would break their hearts to be parted for another day.) The couple were charmed.

'These are so beautiful dogs,' the girl cooed. 'Please, what make are they?'

The couple were, I thought, German. Over the last few years, German visitors have been coming here in ever-increasing numbers. I said the dogs were Golden Retrievers, and explained how the breed had been developed to produce a good-looking, even-tempered gun-dog, how their superior scenting ability was the result of a one-off cross with a bloodhound many years ago. I didn't bother to mention that, in Tom's case, this genetic engineering was wasted because since puppyhood he had remained convinced that socks, which he will go to great lengths to retrieve from wellies, are actually pheasants.

The couple listened with a rapt attention that was entirely gratifying. Before I knew it, they had sat down on the grass beside me, and were asking about my life here, where and how I lived. I explained as best I could.

They were as alike as two peas in a pod. Both had straight, silky blond hair and startlingly blue eyes and, apart from the tender glances they constantly exchanged, they could have been brother and sister.

'You are most fortunate to live in such a beautiful place,' the young man said gravely, waving a sun-tanned hand to indicate the surrounding splendour. 'In Germany now the air is not clean, the water cannot be drunk and everywhere is so crowded. Not like this. I did not believe there was anywhere like this.'

We all fell silent for a while, contemplating the wisdom of his words. He was right of course. I was very lucky. I had a sudden vision of Los Angeles, and how it would be there at this time of year – choking smog, densely-packed humanity all scurrying around trying to be thin, rich and famous, preferably in that order, and the miles upon unrelenting miles of concrete that seemed to have sucked the very soul out of the place. How on earth could I have let a little set-back like electricity make me lose sight of all that I had now? There was a way round the problem. There had to be.

The young couple probably weren't sure why I thanked them so profusely as we said our goodbyes. Just before they disappeared down the forest road, they turned and waved enthusiastically. I thrust the fateful letter into my jeans pocket.

By the time I had walked the quarter of a mile back up the boreen to the house, I could hear the small camping kettle whistling dementedly in the kitchen. Kitty still slept peacefully and the rendering still clung grimly to the walls waiting for me. I had a quick cup of tea and went outside to tackle it.

Soon my hatchet was rising and falling in a steady rhythm. Rendering flew everywhere and great billowing clouds of lime mortar dust smothered me. As I worked, I considered the problem. Lickeen was, or had been, a farm. Surely there were grants available for farms, farmers in need of electricity? Before massive depopulation had decimated this part of rural Ireland, I had been told that three strong men had worked this place, brothers who rarely went to the village but spent their lives up here in solitary splendour, working their land and raising prize-winning bullocks famed as far afield as Kerry. My dreams for Lickeen weren't quite so ambitious but still, I badly wanted to see the old place restored to something resembling its former glory.

I would investigate the possibility of a grant. Perhaps there was help available that might reduce the terrifying quote to something which didn't so closely resemble the National Debt. Meanwhile, I would just have to manage as best I could without electricity. The builder who was to work with me wouldn't be starting for several weeks yet. There was no time for writing anyway, so forget the word processor, and you didn't need electricity to wield a hatchet.

I uncovered a fair amount of stone that first day, constantly stopping to admire the skill and craftsmanship that had gone into the building of this old house. Those builders hadn't had electricity and look what they had managed to accomplish.

By about three that afternoon, I felt as if I had been arm-wrestling with several grizzly bears. The steady, rhythmic sound of the hatchet hypnotized me. I felt like the girl in *The Red Shoes* who eventually dances helplessly off into the sunset because she can't take off the shoes. I had decided to go back up to Barley Lake Cottage and start phoning around about grants but I couldn't put the hatchet down.

When I finally forced myself to stop, my right arm felt wobbly and strange, and I imagined myself developing cartoon-type muscles like Popeye's during the night. I woke Kitty from what looked like a very good dream, put my things into the backpack and whistled for the dogs. As we headed down the boreen, the first drops of rain spattered onto the ground, making me feel better about quitting work while there were still many hours of daylight left.

Barley Lake Cottage was clean, cool and welcoming. As I sluiced off the worst of the dust and debris under the tap, I was glad the sale hadn't gone through yet. To hell with the bridging loan, I thought, putting the kettle on. Trying to live at Lickeen while I was still working on it would have been pure misery. I settled down at the kitchen table with a mug of tea and a notepad. I had to dial with my left hand because my right wouldn't do anything it was bidden. It felt hot and inflamed and I wondered if I hadn't perhaps overdone things. Outside in the back field, I could hear Kitty and Anlon, happy to be reunited, whickering softly to each other.

I began my marathon bout of phoning by calling Teagasc, the Farm Advisory Service. They were kind and helpful when I explained my predicament. There were such things as farmer's grants for electricity, a sympathetic man assured me, but they were being phased out because most people, even in the more remote areas, had light now. In any event, he added, the people I needed to talk to were at the Department of Energy. He gave me the number.

I quickly plunged into a bureaucratic nightmare, shunted from department to department, kept on hold for ever, and inexplicably cut off mid-flow. A succession of bemused clerks tried to assimilate the fact that I was actually 'Denise Hall', a woman, not 'Dennis', a man, that I had bought a farm which I intended to work, and that I didn't have electricity which I very badly needed. None of them were unhelpful – it was just that my whole situation seemed to throw them into a panic, a response with which I could entirely sympathize. It terrified me as well. It was nearly five o'clock by the time I found somebody who didn't want to palm me off onto somebody else. He told me what I had to do, wished me luck and said I could call any time if I needed more information.

I hung up the phone and stretched my weary body out on the couch to read the notes I'd made. I had been scribbling furiously with my left hand as he had been talking and my writing was almost indecipherable. Could I have got this wrong? According to my notes, I had to get a herd number in order to be eligible for this grant. To do that I had, quite reasonably, to get livestock. The going rate seemed to be one cow or ten sheep. Kitty did not qualify as livestock although the friendly man in the Department had been unable to explain why. The grant was generous and would reduce the cost from the impossible five thousand to around fourteen hundred. It seemed that, like it or not, I was about to become a farmer and there wasn't much I could do about it, not if I wanted electricity.

Gingerly, I flexed the aching muscles of my right arm and reached over to turn on the television. The sappy tones of *Home and Away* filled my small living-room. It seemed like the perfect antidote to a day full of render-bashing and financial crisis.

Outside, the mountain turned a hazy blue in the long, slow dusk, promising good weather the next day. Ewes called to their lambs. My neighbour passed by with his small herd of cows, bringing them down off the mountain as he did every night at about this time. I was only dimly aware of all this activity. The soporific effect of Australian accents and comfortably predictable story-lines quickly lulled me into restful slumber.

I woke several hours later, cramped and uncomfortable. Although it was only nine thirty, it felt like two. I let the dogs out for a last look around, made a half-hearted attempt at tidying up, then climbed the steep stairs to my small attic bedroom. I was under the covers and asleep again in record time.

That night, cows and sheep dominated my dreams, huge livestock markets full of them and me wandering in between the crowded, smelly pens, trying to avoid the appeal in their unfathomable eyes. Then my dream suddenly changed and I became the solitary passenger on a train that appeared to have no driver and was hurtling along improbably narrow tracks at break-neck speed. But I felt strangely calm, relaxed, knowing somehow that the only thing to do was to sit back and enjoy the ride . . .

CHAPTER SIX

ENTER ELECTRA

Dawn next morning matched my own mood perfectly – grey and sluggish. Rising at absurd hours has become a habit with me. It started, as I suspect it did for many women, when my daughter Amber was a baby. Since then I've woken at first light whether I needed to or not.

I lay in bed, staring out of my small window at the thick mist which blotted out the mountains, and flexed my right arm tentatively. To my surprise, apart from a slight twinge, it didn't feel too bad. I wished I could have said the same for the rest of my body. It was definitely not happy, and wanted to stay right where it was for at least the next couple of years. Despite my stern reminders that there was work to be done, it buried itself under the duvet and stubbornly refused to budge.

By the time I made it down the stairs half an hour later, the mists had been replaced by a fine rain and the cottage was filled with thin shafts of watery sunlight. That meant rainbows. I filled the pint mug I use for the vital morning coffee and shuffled outside to have a look.

A stunning double rainbow arced gracefully over the

glen, its end buried somewhere in the mile-distant lands of Lickeen, suddenly visible again from Barley Lake Cottage. I sat on the picnic bench outside my front door, sipping coffee and gazing at the display.

I've always been a sucker for omens, but never more so than since I bought Lickeen, when I've felt in particular need of any reassurance I could get, real or imagined. I watched spellbound as the rainbow dissolved then re-formed again in the slowly strengthening light, its colours even more brilliant than before. It was going to be a good day after all.

By half past eight the sandwiches were made and the tools I would need that day assembled. Hatchet (of course – already that small implement had begun to feel like an appendage), slasher, because you never knew when you might need one, and a bow-saw so I could start cutting some firewood. But, first, I had one quick phone call to make.

While rainbow-watching, I'd come to a decision. I would buy a cow. I've never been particularly fond of sheep so the choice was not a difficult one. For some time, I'd harboured a secret hankering for a Kerry cow, small, sleek, black creatures, which used to proliferate in these parts when durability and a wide-ranging appetite were more important than maximum milk yield. For this important step, I needed to call on the help of the professionals, farmer friends of mine from a nearby village, who with a combination of wit and humour had talked me through the trauma of Kitty's first foaling.

By now my friends would have just about finished milking their own well-run herd of Friesians. Dermot and Helen didn't even flinch when I told them I had to buy a cow right away, or when I said I rather fancied a Kerry, whose prices have increased dramatically since they became rarer. They were off to the cattle market anyway, Dermot said, and they would see what they could do.

The rain persisted for most of the day. It was so humid that whenever I put on my waterproof jacket, I quickly became wetter than when I wasn't wearing it. In the end, I worked doggedly on up at Lickeen without it, my T-shirt clinging damply and my hair plastered against my head. After the first hour I didn't notice the stiffness in my arm any more, and my work assumed the same hypnotic rhythm as the day before. By the time I was ready to stop for lunch, I'd uncovered a respectable amount of stone and I was beginning to think that by the time my impromptu family arrived in September, I would probably be in residence. There was really nothing to it, I thought smugly, managing somehow to ignore the rest of the dereliction and lack of facilities that would still exist after the render was no more.

I sat under one of the huge old sycamores with the dogs and ate my sandwiches, sheltering beneath its generous branches and listening to the steady drip, drip of rain on leaves, the glen spread out beneath me in soft-focused splendour, full of diffused greens and dark smudges of purple. Lickeen had a dream-like quality about it today that I found entirely mesmerizing, if a little soggy.

The next morning, Dermot called. He was at the cattle market again. Not that he needed to tell me. The noise in the background was incredible. 'I think I've found a Kerry for you,' he shouted, in a brave attempt to make himself heard above the bellowing. 'The only thing is,' a hesitant note entered his voice, 'there's a bit of a problem with her.'

'What's that?' I realized I was shouting now too.

'Well, she's only got three tits.'

He sounded distinctly apologetic, as though this unfortunate disability was somehow his fault. I pictured the poor animal now, bewildered by the noise and confusion of the market and probably the object of many a ribald joke from farmers less sensitive than my friend.

'On the other hand,' he roared, above what sounded like the squealings of an enraged boar, 'I should be able to get her for a handy price. That's if you don't mind about the three tits,' he added hastily.

I assured him that I was not the sort of person to let mere physical appearance stand in the way of a promising relationship. Dermot explained that although the poor beast had lost one teat to mastitis, she had fully recovered and it hadn't affected her milk yield one bit.

I spent the rest of the day hacking rendering and wondering if I would be the proud owner of a three-titted Kerry cow by evening. I was in the back field at Barley Lake Cottage brushing the horses when the phone rang. It was my personal hotline to the cattle world.

'Your man was asking too much for her,' Dermot said without preamble as soon as I picked up the phone. I was still gasping for breath from the dash to the house. 'And I couldn't get him to see sense, so I didn't buy her after all.' I knew my friend was a master at the delicate and skilful negotiations involved in buying and selling cattle. If he said the price was too high, then I wasn't going to argue with him. 'She might be a pedigree Kerry but, in my book, a cow with three tits is a cow with three tits and never mind the papers,' he concluded firmly.

There seemed to be no answer to that, or not one I could readily think of, so I simply thanked him for trying. Dermot said he would keep looking. The right cow at the right price would turn up sooner or later.

By the end of that week, I wasn't giving much thought to cows. The adrenalin rush which seemed to have sustained me through those first hectic days had subsided and I felt tired and cranky. When I looked at the wall I was working on, I didn't see progress or even potential. I didn't think, a few more weeks then on to the inside, neither could I imagine roses clustered around the doorway (when I finally got one) or smoke drifting out of a rebuilt chimney-stack, or any of the other visions which had formerly sustained me.

When Dermot called back with more news on the cow front, I had almost forgotten I was supposed to be getting one. I was much more concerned with smoothing layers of

Vaseline onto my blistered hands and entertaining fond thoughts of life in a nice quiet city.

'I've found just the cow for you,' he said cheerfully. 'A nice little red dairy Shorthorn. The prices on Kerrys are a lot higher than you said you wanted to pay. I happened to spot this one in a field as I was driving past, and I thought I might as well see if she could be bought.' This sort of challenge was life-blood to my friend.

Remembering that without a cow there was no hope of ever getting electricity at Lickeen, I put down the jar of Vaseline and paid closer attention.

'She looked perfect so I went ahead and bought her,' he was saying. 'If you're going to be in tomorrow, late afternoon, say, I could deliver her then.'

I said I would, thinking sneakily that at least I would have a legitimate excuse for knocking off work at Lickeen earlier than usual.

There is a short but improbably steep hill leading up to Barley Lake Cottage, so it was easy enough to hear my friend's Land-Rover, trailer attached, grinding its way laboriously towards my house. Dermot backed the vehicle expertly into my small parking space and let down the ramp. The little red heifer came out at a fast clip. She hurtled down the hill to the bottom of the field where she stood glaring at us, snorting occasionally, remembering, no doubt, the recent indignities she had been subjected to. She seemed distinctly unimpressed by her surroundings. I decided to leave her to it for a while and went inside to make some tea and settle up with my friend. Besides, I wanted to hear the story of how she had been bought.

It was a good one, full of the usual fancy footwork and smooth talking that characterizes such encounters. Dermot had managed to buy my heifer for four hundred pounds, a very good price. When I offered an additional payment for transporting her, Dermot declined, saying it wasn't far out of his way. If there was anything I needed to know about caring for my cow, I should call him, he said, leaving with a cheery wave followed by a dramatic crashing of gears as he negotiated the short, steep hill.

The little red heifer stood in a corner of the field and sulked. When I tried to win her over with a few dairy nuts, she turned her back on me pointedly. After making sure that she had a full bucket of water, I left her to settle in at her own pace.

She was only a little cow but she had a very big voice. In the middle of the night I struggled through layers of sleep into semi-consciousness, convinced for a moment that I had a lovesick moose outside the bedroom window. Inexperienced as I was, I knew that she was lonely, confused and wanted to be sure that the whole world knew the depths of her despair. Soon every other cow in the glen seemed to be responding to her frantic SOS. The din was unbelievable and I knew I wasn't going to be very popular with my neighbours the next day.

I amused myself by trying to imagine what the other cows were saying to her as they roared back and forth from field to field – something along the lines of, 'Don't worry, dear, you'll get used to it, and tomorrow perhaps you can break out for a bit of a chat with the rest of the girls.' Then I began to feel guilty at having deprived my cow of the company of her own kind. She was obviously a sensitive animal. Perhaps I should get up and make her a bed on the porch, where she'd feel a little less lonely? No, I was going to have to harden my heart and cover my ears before I did something rash. She would just have to get used to it. As the bellowing increased in volume, I finally fell back into a deep and dreamless sleep.

Early next morning, clutching another handful of dairy nuts, I went down to visit my vocal little animal. She looked entirely innocent, as though she had spent a peaceful night in quiet slumber. She stood her ground until I was nearly up to her, then snorting suspiciously she raced away to the far side of the field, her tail held high in the air.

But this boorish attitude did not persist. Over the next few weeks she underwent a complete personality change. She was christened – rather wittily I thought – Electra, by a neighbour who had fallen for her doe-eyed charm, even though she, too, had been kept awake that first night. Elly, as she became known to her friends, soon accepted her solo status with considerable enthusiasm. In fact, I had the uncomfortable feeling that, not having other cows to identify with, she actually believed she was a dog. She trotted promptly up to me when I whistled. She demanded that large amounts of time be spent scratching her under the chin and other hard-to-reach spots. Now she sported a rather nifty brown head-collar recently outgrown by Anlon, the foal.

Since I now had a cow, it seemed only logical that, as well as making electricity possible eventually, she could also provide me with milk, butter and cheese. But the miracle of milk was not going to happen of its own accord. Even I knew that. Over the months I had come to admire our local vet, Finbar, and I decided to get him to come out and take a look at her. Daily he faced very large animals who were often less than pleased to see him. Beneath his necessarily tough exterior I'd noticed a care and concern that I had come to rely on.

When Finbar arrived, I was eating brown bread and boiled eggs, too tired to cook a proper meal. I hadn't even bothered to have a wash after the day's work. Not that Finbar cared about that. After testing cattle for TB in the fields all day, he was covered in mud and cow dung, and there were streaks of blood down the sides of his trousers.

It didn't take us long to get Elly herded into a small shed at the bottom of the garden. Finbar quickly had her in a deft headlock. He completed the necessary internal examination to see if she was mature enough to produce a calf, ignoring

Elly's bellows of outrage. When she roared, she looked straight at me and I got the feeling that she was expecting me to help her out somehow. So I pretended not to notice.

The vet withdrew a mucky arm from somewhere deep in Elly's nether regions and peeled off the thin plastic glove he was wearing.

'You won't be needing the AI man anyway,' he said cheerfully, lighting another cigarette and giving Elly, who was still snorting and trembling indignantly, an affectionate pat on the rump. 'This heifer's about a month gone.'

'She's what?' I heard myself croak as if from a great distance.

'About a month gone, give or take a day or two,' he repeated, obviously amused by my response.

This was not at all what I'd had in mind. I had imagined a planned, clinically controlled pregnancy – the white-coated AI man bearing the fruits of the bull of your choice in a frozen straw, that sort of thing. This calf, however, would be the result of a moment's madness, combined with a bit of judicious fence-hopping on someone's part. Its sire was unknown. Elly remained smugly silent on the topic. Perhaps I only imagined it, but as I was letting her back into her field I could have sworn that she had a particularly jaunty, devil-may-care bounce to her walk.

I said goodbye to Finbar and thanked him for coming out.

If only she could have waited, I thought peevishly, visions of calving jacks and ropes, none of which I possessed, flooding my mind. Since the AI man had taken over where nature left off, there are few bulls around in the fields any more. Those that are tend to be large, for breeding beef. That meant there was a better-than-average chance Elly could have a calf which was too big for her small frame.

When this was all over and I finally had electricity, the simple act of turning on a light switch was never going to seem quite the same again.

CHAPTER SEVEN

THE OLD WAYS

When my new home at Lickeen had been built a hundred and eighty years ago, electricity had not been a consideration. The most important tools were not tractors, diggers or cement. They were muscle-power and a good eye. As I continued to hack away laboriously at the rendering, I became more and more curious about those early builders, how they had ever managed to raise the enormous corner-stones that now stood majestically exposed at each end of the house.

As is often the way round here, a neighbour friend of mine had the answer. I met him early one morning at the gateway to Lickeen where I parked the car since the boreen was still undriveable. I was going through the daily procedure of sorting out what I would need for work from amongst the car's chaotic contents and trying to decide whether to face my arch-enemy the strimmer, and do an hour's clearing while I still had the energy. I had become convinced that this machine hated me. It purred into life after only a few pulls for anyone else, but if I yanked on the cord, it required at least thirty tugs before sullenly spluttering then promptly cutting out again.

41

'A dry morning so far, thank God.' My neighbour Jimmy's voice cut through these musings, gravelly and unexpected in the early-morning stillness. He stood behind the car grinning amiably, his black and white collie sitting attentively beside him.

We talked for a while about the progress up at Lickeen. I told him about the vet discovering that the little red heifer was already in calf. Jimmy said I shouldn't worry, that everything would probably be fine and, anyway, I could always call on him if I needed help. As we were chatting, I remembered that Jimmy had worked with stone himself once. He was retired now, but he would probably know how they had raised the massive corner-stones at Lickeen.

When I asked him he looked at me thoughtfully for a moment, pushed his battered cap further back on his head and ran his fingers through his steel-grey hair before settling himself comfortably on the top rung of the gate.

'I'll tell you how they did it. They did it because they were mighty men and we'll never see their like again. You couldn't get three men together to work on a place today, let alone the twenty or more who probably helped with your house,' he said vehemently. 'Most of them have left now, gone to America or England, but it wasn't like that round here when your place was built. Neighbours would have got together and raised those big stones with some sort of winch they would have built.' He lapsed into thoughtful silence.

Ireland has suffered from successive waves of emigration over the last hundred and fifty years. Famine, colonialism and lack of employment have all taken their toll. West Cork was particularly hard hit by these ills and many of the men who were Jimmy's contemporaries left a hard and difficult way of life years ago for the comforts of the city and the promise of a weekly pay-packet. For those who stayed behind, wives became hard to find. Many women no longer wanted to marry a subsistence farmer working thin, rocky soil and living a life that, despite its quiet beauty, was filled with hard work and isolation.

'I suppose you'll be plastering over the old stones again soon?' Jimmy asked eventually.

'I won't,' I answered, horrified at the thought. 'They're perfect, so I'm going to leave them just as they are.'

A broad grin lit his craggy features. 'You're right, too. I've always thought it was a funny thing how we became ashamed of the stone, started covering it up. After all that trouble people must have gone to getting the old places built, gathering the stones off the land, matching them up, shaping them and all that. And for what? So that a shower of eejits could come along and cover them all up again.' He shook his head disgustedly.

Yet it wasn't only the old stonework that had been affected by changing tastes and times. Skilfully woven thatches were ripped apart and replaced by sheets of corrugated iron, and huge old fireplaces, big enough to roast oxen in, were bricked up, their majestic lintels lost forever. There are plenty of people, though, who feel no rush of nostalgia for the old days, the old ways. For them, the past is full of dark shadows. Today they want comfort, convenience, modernity, and they do not regret the passing of the old thatched cabin and a simpler way of life at all.

I became aware that my neighbour was reaching into his voluminous pockets for something. He pulled out a bottle. 'We'll have a drink to the old ones,' he said firmly, wiping its top and taking a fair-sized swig himself.

Soon, we were passing the bottle backwards and forwards companionably, talking of stone, the incomprehensible behaviour of some tourists and, as far as I remember, the meaning of life.

'Yes, they were mighty men in those days all right,' Jimmy said, switching back abruptly to his former topic.

I agreed, or tried to, but the words seemed to stick to my thickening tongue. I was, I realized with some alarm, more than a little drunk and it wasn't even nine o'clock. I have always prided myself on being able to keep up with the best of them when occasion demands, one of the many dubious skills I acquired during my twenty-odd years as a journalist. This morning's session was, however, a little rich for my blood. I slid carefully off the gate and mumbled goodbye to Jimmy, who, I noticed with some annoyance, still looked perfectly composed. To my surprise, by the time I had made my way unsteadily up the boreen to the house, I felt terrific, ready for anything.

I was certainly capable of tackling a little undergrowth. I grabbed the strimmer by its scrawny neck. I gave its irritating cord one sharp, no-nonsense sort of pull and for the first and only time since I have owned it, the wretched thing roared obligingly into life. I worked away happily on a dense patch of brambles for the next hour, thinking dimly that there was probably a moral in this tale. Perhaps it was just as well I was unable to grasp it.

Over the last four weeks I had become so used to working alone up at Lickeen that the day I first heard the throaty growl of the builder's tractor I almost jumped out of my skin. I watched the top of Ray O'Shea's venerable blue tractor appear over the top of the fuchsia bushes with some relief. Staggering the quarter of a mile up to the house carrying heavy loads of one kind or another had lost its limited charm long ago. Now we would start to see some real progress. The boreen was still impassable to anything but heavy-duty vehicles, although work was due to start on it soon. The JCB driver who would be tackling the job had looked it over and we had agreed on a price. Now it was just a question of waiting for him to fit us in.

Ray leapt down and stood beside his machine which continued to roar like some shaggy, good-natured beast. I noticed there were bags of cement and a pile of sand in the trailer.

'I'll just dump these and go down for more. Then we can get started,' Ray said cheerfully.

Soon, cement was neatly stacked inside the house to keep dry, and there was a huge pile of sand round at the back. There seemed to be a new sense of purpose about the place, a professionalism that had not been achieved by me and my hatchet. Ray whistled cheerfully as he unloaded supplies. When he turned off the tractor he stuck an old welly over the funnel to keep its innards protected from the frequent showers we were having that day. The boot gave the ancient machine a rakish air which quite became it.

'I think I'll make a start on the new window,' he declared. Still whistling, he disappeared inside the house. In a short time, he had knocked a sizeable hole in the kitchen wall where we had decided to put an extra window. An unaccustomed light began to filter into the old, cobwebby room.

I had rendering to knock. Sometimes I thought I would always have rendering to knock. But I couldn't seem to buckle down to it today. This unusual activity inside the house had set my mind whirring. I wandered round its decrepit rooms, seeing fires blazing in the two hearths, beams cleared of the thick, sooty tar which coated them, walls free of the cracked and yellowing plaster and freshly painted in vibrant colours.

Each room seemed to have a different atmosphere, which I liked. The front room, just off the kitchen, with its small Victorian fireplace was shut away from the rest of the house, a room you could hide in at the end of the day and read books, watch television, although it took a great effort of the imagination to picture that at the moment. This room and one of the two bedrooms upstairs looked the most dilapidated. Great chunks of plaster had fallen onto the floor in both, and their windows were no more. The room that I had picked to sleep and work in wasn't quite as bad. I particularly liked the huge A-frame joists that were, thank God, still intact.

Although the kitchen was dingy and discoloured now, you could tell that it had once been the heart of what seemed to have been a happy house. None of the rooms were what you might call large and the ceilings were low, but the house was more than big enough for me. I loved it. And somewhere upstairs there would be enough space to build a small bathroom, my first in three years. I forced myself to stop daydreaming and got back to reality.

It was good to be working with somebody else, particularly someone who knew what they were doing. I did more than usual that morning, ignoring the frequent fierce showers, my head full of plans for transforming the old place. By dinner-time, Ray had erected a rough frame for the new window, and he left for a while to get something to eat. The kitchen had even more rubble in it now and it was full of dust from where he had knocked the hole in the wall. But I was glad to be out of the rain. Over the last few days, the weather had been terrible and Lickeen had turned into a sea of mud.

I rubbed my sopping hair with an old towel, remembering how, before we owned Lickeen, when summer still seemed a possibility, I had pictured myself tanned and fit, rebuilding the old place single-handed, laughing at the rapidly forming callouses, tossing bags of cement about with light-hearted ease. Inspired by this heroic vision, I had driven into Bantry and bought a rather fetching pair of black Lycra cycling shorts and a tube of sun-tan cream. It had rained ever since.

I searched in my backpack for the apple I knew was in there somewhere. Water from my still-wet hair dripped steadily down the back of my neck. Instead of the apple, I found the once-stylish shorts. They had sprouted a healthy-looking growth of greenish-grey mould. Further exploration revealed that the sun-tan cream had been leaking out of its tube and had formed a gelatinous mess across the bottom of my pack. I cleaned up the cream and threw away the shorts. Then I reached for the list I carry around to help me with the weekly shop. I added 'Thick Socks' to it and, as an after-thought, 'New Wellies'. Now let's see what the weather will do, I thought crossly.

Despite the rain, Lickeen began to look and sound like a real building site. There were lengths of sweetly scented pine and rolls of roofing felt up at the house now as well as sand and cement, and when I went home at night, it was with a new sense of satisfaction. In a few days, Ray would be starting on the roof. He would build a new chimney-stack: the cracked and dangerous old one was the first thing I glanced at every morning, half expecting to find that it had collapsed during the night. Then he would carefully remove all the old slates, as many of them could be reused, lay roofing felt, and then put the slates back on. I didn't want to think about what would happen if it rained while the old house was exposed to the elements.

In another week or so, I would have finished removing the rendering from the outside walls. It would be time to start pointing between the many stones with a mixture of lime and cement. So far, we had managed well enough with the odd bucket of water from the stream, but now that there was building work to be done we would need a supply close at hand. Ray said it was a question of finding a likely spot in the stream above the house, anchoring a pipe in one of its deeper pools and hoping we could get the water flowing by gravity. I had serious reservations about this project. I couldn't imagine it was going to be that simple. No pumps, no drilling, just a couple of rolls of piping? Ray had no such qualms. 'Sure, it'll be fine,' he kept telling me.

One morning he wandered off up the mountain with the pipe, a breeze block and a slasher. When I saw him again, he was trailing a length of pipe in one hand and grinning triumphantly. Water poured from it. 'It's all done,' he said nonchalantly. 'Now I'm just going to see if there's enough pressure.' Still holding the pipe, he climbed up the ladder onto the roof, held it above his head and still the water flowed. It really was going to work.

Now we had water right outside the back door. To start the flow, you simply removed the twig that was stuffed into the pipe and hey presto, there it was. The end of the pipe which was lodged in the stream had a small filter attached to it, but, apart from that, the water was completely unadulterated. It was the coldest sweetest water I had ever tasted. Things were definitely looking up.

CHAPTER EIGHT

THE ACCIDENTAL FARMER

Caring for animals had become a major part of my life. Since I had no previous background in farming, I was having to learn as best I could from books, neighbours and hands-on experience. Every day was a new adventure.

Anlon and Kitty were back in their field in the forest enjoying the attention they were receiving from the influx of summer visitors to the area. Elly was grazing at a neighbour's. Each evening after I had finished working at Lickeen, I would set out with Tom and Sam to feed and water them.

One evening I worked late, determined to get a particular patch of wall finished before I went home. Ray O'Shea had roared off on his tractor some time ago. Apart from the rise and fall of my hatchet and the evening chorus of birdsong, Lickeen was still, silent – too silent. Dark, ominous-looking storm clouds were forming overhead and I still had the livestock to see to. This was no kind of weather for the end of July. Hurriedly I stuffed my belongings into the backpack.

By the time I had seen to Kitty and Anlon, the first fat drops of rain had started to fall and the wind was howling. When I reached Elly's field near Barley Lake Cottage, it was pouring and I could feel the inevitable layer of lime-mortar dust slowly stiffening on my body. It was dark and the torch, as usual, was not working, so I picked my way cautiously across Elly's field, whistling for her as I went. There was no response. I dumped her food unceremoniously out of the bucket onto the ground and headed for home.

I couldn't have gone more than a hundred feet when I heard the pounding of hoofs behind me. Elly was doing a passable imitation of a Spanish bull about to have her moment of glory with the matador. Head down, she snorted and pawed menacingly at the ground. It was obvious that she had run right by her food and intended to flatten me at the first opportunity for attempting to leave the field without feeding her. When it comes to her dairy nuts, Elly is not a girl to be trifled with.

I whirled the empty bucket over my head a few times for dramatic effect and shouted, 'Look, you silly cow, it's empty – see, your food's on the ground over there.'

Elly was not impressed, and charged me again. Trying to run backwards over old potato ridges in a dark and muddy field is not a good idea. If I didn't want to be floored by several tons of angry heifer, I realized I was going to have to make a stand.

I wasn't really sure what to do next. None of my reading had prepared me for this sort of eventuality, so I bunched up my right fist, considerably strengthened from several weeks of render-bashing, and landed Elly a fair belt on the jaw. I don't know which of us was the most surprised. Elly skidded to a halt, staggered slightly and glowered at me. For a few seconds we both stood stock still, eyeing each other suspiciously. I wondered if she was getting ready to charge again. What would I do then? A full body-slam this time, just to show I meant business?

Then, with a look of utter contempt, Elly flicked her tail a couple of times as if ridding herself of a particularly annoying fly, sniffed, turned on her heel and stalked off. When I considered it safe, I beat a hasty retreat towards the gate, hoping she wouldn't hold this unfortunate incident against me. How long *was* a cow's memory? If it was revenge Elly was after she would get her chance soon enough because, after she had calved, I was going to have to learn to milk her.

Thanks to Elly, I had already filled in the first flurry of forms for my electricity grant and had recently received notification that I now had a herd number. This entry into the farming world had been rapidly followed by a visit from the Department.

Their representative had arrived at my door during a fierce downpour, resplendent in suit, tie and clipboard, and politely requested to inspect my herd. This did not take him very long. He studied Elly seriously for a few moments while she glared back at him. When this inspection was over, we went inside to drink tea and dry off. Despite my herd of one, he appeared to take my farming aspirations quite seriously, and, during the next half-hour, gave me lots of useful advice.

I discussed my plans for Lickeen in depth with him that day and by the time he left they didn't seem quite so far-fetched. Eventually, I wanted to move the Sitka spruce from the pastures on which they had been planted, and reclaim the land, so that I could have more cattle and horses. The Sitka could go further up the mountain where it was rougher, which, in my opinion, was where they should have been planted in the first place.

In this glen, where there is precious little grazing because the soil is thin and rocky, it is heartbreaking to see good grass ruined in this way, but as small farms have continued to fold, the ubiquitous grant-aided Sitka often claims the land after its people have left.

Apart from Sitka, Lickeen was rich with a wide variety of broad-leaf species – ash, beech, oak and holly. These trees would need managing, caring for and protecting from the encroaching rhododendrons which threatened several areas of native woodland in these parts. My land was infested with them too. That meant hacking, burning and clearing so that sapling trees could have the space and light to grow.

The lands of Lickeen would provide work for several lifetimes. The farm had been fashioned slowly, over several hundred years, and the work put into it must have been backbreaking. Now that slow, painstaking process would have to begin all over again. But, today, there were tractors, electricity (I hoped!) and JCBs to help me accomplish what I dreamed of.

When I first saw Lickeen, I found myself gripped by an entirely atavistic desire to own land, land that I could pass on to my child. Not just any land, mind you, but this particular wild and lovely mountain. There was magic here – there had to be, because already Lickeen had caused me to perform feats of which I had not formerly considered myself capable.

A few weeks after I had been granted the herd number, and had been inspected, strange and often incomprehensible leaflets began to appear in my post – EC directives on fertilizers, cautionary tales about the dangers of the bot-fly worm and Mad Cow disease. It was all a bit bewildering, but at least my grant application was well along the pipeline. Now it was just a matter of waiting to hear if I would be approved.

I was sharing an afternoon tea-break with Ray O'Shea, talking about what we still had to do and how much easier several of the tasks would be with power to help us, when a neighbour and her four-year-old son dropped by for a visit. He soon got bored with the uninteresting adult conversation and wandered off to amuse himself.

My neighbour was telling us about some of the difficulties she and her husband had experienced when they were building their house several years ago. Suddenly she stopped mid-sentence and looked around her warily in the way that

'The farmhouse at Lickeen was perched half way up a mountain and derelict, the track to it undriveable. There was no water or electricity and it was obviously going to take massive amounts of time and money to make the place habitable. . . I wanted to live there very much.'

Above: 'I had it on reliable information that a hatchet was the best implement for knocking off the many tons of rendering. . .'

Left: It required a huge effort of the imagination to picture the future bedroom and kitchen.

'As long as there was grass I knew that Kitty would stay close to the house.'

Right: Anlon, uncharacteristically quiet, dozing in the sun.

Barley Lake Cottage for sale.

Electra, bought out of necessity to provide electricity for Lickeen, views her world through regal eyes.

As autumn fades into winter, work
on the outside of the house finally
draws to an end – Tom and Sam are
clearly proud of the results.

The bliss of parenthood.

Inside, the house comes to life once more and Denise achieves her first bathroom in three years.

Top: Hard at work in the kitchen – Denise. . .

Left: and Kitty.

Below: Life in West Cork.

mothers will when things go quiet and their children cannot immediately be seen. 'He was here just a minute ago,' she said, rising to her feet and looking around apprehensively. We all spotted him at the same moment in a field by the side of the house, busily engaged in milking Kitty, who was at Lickeen doing some land reclamation work of her own.

Kitty's udder was full and fat. Farmer's lad that he is, my neighbour's son had seen the teats hanging around doing nothing, and had decided that he may as well make himself useful by doing a spot of much-needed milking.

His mother disentangled her son from Kitty's teats. A stream of dark yellow milk was flowing nicely. She explained that Kitty wasn't a cow and would be needing all her milk for the new foal. Her son looked distinctly disappointed. Kitty didn't seem all that bothered by this indignity though. She simply turned her large head and watched the young lad being led away, then went back to switching her thick black tail to ward off the flies. I sometimes get the distinct impression that Kitty, in her wisdom, has long since passed the point where she is surprised by anything our strange and incomprehensible species might do.

After my visitors had left and Ray had gone back to his current project of dismantling the roof, I stood and watched Kitty for a while as she contentedly munched. She looked as full and as ripe as a summer fruit herself. I was beginning to look forward to the birth of this new foal with a keen sense of anticipation.

There were, as the blacksmith often reminded me, few mares like her left – large, honest beasts, hard-working and good-tempered but often with a touch of thoroughbred blood in their veins. The much-maligned workhorse was becoming obsolete, its hairy feet despised as a sign of its peasant origins, and its tremendous strength no longer valued. Now that I knew what it cost to keep a horse her size over the winter, I could see how this had happened. Kitty's appetite was enormous. Like Lickeen, she had been a commitment I had taken on blindly, unable to know what it was I was letting myself in for, but propelled by the lemming-like instinct that seemed to have overtaken me since I moved here, to plunge, with reckless abandon, into unknown territories.

When things had calmed down a bit, I intended to try to find a set of working tackle for Kitty, the harness and traces that would enable her to plough and drag logs out of otherwise inaccessible spots. On land like this, where fields are small and stones plentiful, there is much to be said for working with a horse. They can get to places no tractor ever could and can turn over the earth right to the edges of a small field. I didn't have a tractor anyway, but I did have Kitty who had been used to working all her life. In her maturity, occasional light duties would be good for her and help to keep her fit. Her foal was due in about another month and after Kitty had recuperated I would find somebody to teach me how to work her. I rooted in my pocket for an apple core I had been saving and called Kitty over, holding it out on the flat of my palm so she could lift it delicately with her thick lips. Once I got the knack, working with Kitty would be a rare pleasure.

Seeing Ray perched up on the roof made me feel suddenly guilty. I should be working too, instead of standing here wool-gathering. I was trying to get as much accomplished as I could before the family turned up in early September. In that first rush of enthusiasm when I started work at Lickeen, render seemed to fly off the walls thick and fast. I had written to Amber, blithely assuring her that I would be able to move in by August, and Lickeen would be ready for them when they arrived. I realized now that it had been a rash and over-optimistic promise. Work on Lickeen was progressing, but it was slow and laborious and everything was taking much longer than I had thought it would. There were many unforeseen delays – bad weather, supplies that didn't arrive when they were supposed to, and projects that took much longer than estimated to complete.

I sat down to write another letter, a more realistic one this time, telling Amber that, after all, I might not be moved in to Lickeen by the time they came. As I was putting the letter in the envelope, I was struck by a sudden, horrible thought. The proposed completion date on the sale of Barley Lake Cottage was late August. If Lickeen wasn't ready by then, and there didn't seem any chance now that it would be, what was I going to do? I had bleak visions of me and the animals camped out in the cold dark kitchen and the family having a thoroughly miserable rubble-infested holiday.

For the next few days, my mood see-sawed wildly between optimism and despair. The more I did, the more I became aware of the enormous amount of work that still remained. If I was not careful, any sense of accomplishment I felt after a day's work was wiped out by this destructive and pointless approach. I was worried about finances again too. Under present conditions, it was almost impossible to write, and no writing meant no dairy-nuts. I was going to have to find the time to start doing some paying work again, no matter how difficult it seemed.

I had started to point between the stonework on the outside of the house and, although it was slow going, I was finding it a welcome relief from hatchet-work. When the cement dried out to a pale and elegant grey, it highlighted the old stones beautifully and gave me a great feeling of progress.

I was expecting the contractor who would be working on the boreen tomorrow. This would involve JCBs, gravelling and much uprooting of trees as the narrow track would have to be widened. If the weather was anywhere near decent, the contractor thought he could have it driveable within a couple of days. If it rained heavily, he wouldn't be able to touch it again until the ground dried out. The weather was what could best be described as showery, and I had my fingers crossed that it wouldn't get any worse. Apart from the boreen, the slates were off the house and Ray was beginning

Chapter Nine

BOG WOMAN

If it hadn't been that I was driven by visions of having to move out of Barley Lake Cottage in the near future, I would probably have stayed in bed on the day I fell into the hole.

It didn't look as if anyone else in the glen was budging that morning. I looked out of a steamed-up kitchen window at the rain-soaked houses dotted about the hillside, smoke billowing cosily out of their chimneys. Rain was hitting the ground with a stinging fury and the wind howled and moaned around Barley Lake Cottage, but by now I was used to the sudden violence of storms in these parts. It was time to go to work at Lickeen and that was that. Even if I was unable to work on the pointing, there was plenty to be done inside.

The wind wrenched at my clothing and did its best to drive me back in the direction of the house. I should have taken the hint. Instead, I drove down deserted, flooded lanes, past water-logged fields and sodden hedgerows until I got to Lickeen, where, severely hampered by several layers of waterproof clothing, I began the long waddle up the boreen to the house. The dogs gambolled about me as merrily as if it were a perfect spring day. In their book, a walk's a walk and never mind the prevailing winds.

After a couple of minutes, I could hardly see. Rain cascaded into my eyes and down the back of my neck. But I could tell that the boreen was awash. Work had started on it the day before but had had to be abandoned a few hours later. Heavy diggers and mud are not a winning combination.

Great surging streams of water tore down the rough track from high up on the mountain, gouging deep ruts. I peered at this devastation through waterlogged eyelashes. Trees that had been uprooted by the digger lay at forlorn angles, great mounds of earth, dug to be moved from somewhere to somewhere else, had been dumped anywhere when the storm hit. As if all that wasn't enough, several large holes had mysteriously opened up during the night.

Instead of giving in and going back to my cosy dry cottage, I trudged on. Far from the boreen ever being driveable, at this rate it would soon be impassable even on foot. Then the only way to get to the house would, I supposed, be by abseiling down the mountain from the Kenmare Road. And I had imagined I would be moved into this infuriating irresistible house this month. As this bitter thought struck me, I disappeared into one of the larger holes.

It was a deep and particularly wet hole. Immediately, water began to seep over the tops of my wellies. Convinced that I had just dreamed up some new and creative game for their entertainment, the dogs pranced and barked delightedly somewhere above my head.

There was little chance of rescue. No one would be moving on a day like this. I buried my fingers in the mud and began inching myself upwards. This technique meant that my face was squashed into the stuff too. For a terrible moment, all I could think of was the Edgar Allen Poe story about being buried alive. Panting with effort, I crawled up the slippery side of the pit, finally managing to throw myself over the top with what I hoped was commando-style pan-ache. The dogs were impressed at least.

Even though I'd spent the last few months festooned with hay and caked in mud, I still had my standards – hard to detect perhaps, but there all the same; the odd dash of Coco Chanel perfume for no good reason, toenails painted red just because . . . – that sort of thing. But today I knew I had reached new heights in rural *déshabille*. As I staggered through the hole in the wall at Lickeen where the front door should have been, I didn't need a mirror to tell me that I must look like Bog Woman.

Weather conditions inside the house seemed only margin-ally better than those raging outside. The many gaps and crevices that I hadn't pointed yet weren't just letting in floods of rain, they also seemed to be acting like some kind of wind-tunnel. And, of course, the roof was off. Water poured down the stairs and formed muddy pools on the kitchen floor.

I looked around me wild-eyed for something I could do. Then I remembered the wall in the front room. I had been anxious to see how the stone was underneath the decaying plaster, so I had stripped it off one afternoon. When I discovered that it was perfect I thought perhaps I wouldn't plaster inside either. I had pointed the wall and now I could paint it. Painting a wall whose outside has as many holes in it as a Gruyère cheese is not a good idea. I'd go so far as to say that it is a complete waste of time. But by now I was beyond all reason. I painted away enthusiastically while the wind whistled through my hair and my mud-encrusted clothing slowly sealed itself to my body. When I'd finished, I sat on a pile of rubble and admired my handiwork, congratulating myself on persevering.

The next morning dawned fine, clear and, believe it or not, sunny. I hurried up to Lickeen, eager to admire my wall. It looked as if some particularly verminous old drunk had staggered in and thrown up all over it during the night. What

I had thought of as a 'Hint of Lavender' in yesterday's permanent twilight was now revealed to be putrid purple. What was left of it anyway. The rain that had flooded the inside of the house had washed most of the paint away, which was probably just as well. The remnants oozed across the floor in depressing purple streaks. I knew I'd be in for it when Ray saw this masterpiece. It wasn't the first time that my more haphazard working methods had let me in for a good roasting. I thought of trying to pretend I'd been cleaning a paintbrush, but I knew I'd never get away with it. I would just have to come clean, tell him that I had temporarily lost my senses and have done with it. I had no intention of mentioning how I'd fallen into the pit.

The phone's insistent shrill woke me early the next morning. I staggered downstairs to answer it, nearly falling over Tom who was fast asleep in his usual spot at the bottom. 'Hello, here is Rainier, calling from Munich.' The voice sounded tinny and unreal but there was no mistaking the German who was buying Barley Lake Cottage. 'We have just heard from our solicitors in Bantry and they are telling us that everything is now in order. The paperwork that was missing has been found and now we can proceed. So I call to see if the end of August would still be suitable for us to exchange contracts,' Rainier concluded cheerfully.

Suddenly I was wide awake, filled with a heady mixture of delight and panic. On the one hand, it was fantastic that the deal was really going to go through. But where was I going to live? Lickeen would never be ready by then. 'The end of August,' I repeated unnecessarily, trying not to think of the piles of rubble and gaping holes that were my new home. 'That's great news.'

'There's something else my wife and I wanted you to know,' continued my purchaser. 'We will not be able to live there for some time, so, at first, we will just come for holidays. If you would like to stay on at Barley Lake Cottage for a while, until your new home is ready, that would be very fine with us because we wouldn't like it to be empty for too long. It will be Christmas before we are able to return.'

Relief flooded over me. I wouldn't have to share a tent in the field with the animals after all! 'That's very kind of you. I would really appreciate it because I don't think Lickeen will be ready quite as soon as I had thought it would,' I gabbled. 'And the end of August will be just fine to exchange contracts. I hope you are both very happy in this house,' I added with feeling, unable to believe my luck. I had sold my house, the deal was secure, and for the time being, I could continue to live at Barley Lake Cottage.

That evening, I sat down with the bills, the bank statements and a stiff whiskey. It was the moment of truth. I owed the bank a considerable amount of money. Over the last two years, I had not been making enough to live on and had had to take out a series of loans. Then there was the purchase of Lickeen and the cost of renovating the house. As is the way with old places, especially derelict ones, everything cost at least three times as much as I'd thought it would. The point was fast approaching where I would break out in a cold sweat at the sight of another brown envelope.

By the time I had figured out everything I owed, it looked as if I would have about two weeks in credit with the bank, then it would be back to the red zone again. The bills for building materials and work done at Lickeen were going to keep coming for quite some time. I gulped the last of my whiskey and put the paperwork back into my files. If costs kept escalating the way they had been and I didn't make the time to continue writing the stories which had been keeping me and the animals alive, what then? What if, unimaginable thought, I wasn't able to hold on to Lickeen after all? Suddenly I knew beyond a shadow of doubt that whatever it took, I would find a way. I had to. I could never be happy living anywhere else.

The first thing I saw the next morning when I arrived at Lickeen were five brand new windows. Ray had worked late the night before to get them fitted. The frames were teak, and they transformed the house. The two new doors had arrived as well and he was busily engaged in fitting them since it was too wet and slippery to work on the roof. I stood there for quite some time admiring my new improved home. Even better was what nestled in my jeans pocket amongst that day's batch of bills: notification from the Department of Energy that my grant application had finally been approved. The day had definitely got off to a good start.

When I had knocked the plaster off the inside walls, the plumber and the electrician could lay their pipes and cables. After that, somebody, not me, would plaster. I had changed my mind about leaving the stonework inside exposed, although it did look lovely. Ray had convinced me that it would make the house too cold. I had no intention of attempting the work myself – I didn't have the knack and was obviously better suited to unskilled, navvying work.

It had been so long since I'd written any features, I wasn't entirely sure I still knew how. I was certainly having trouble in coming up with any good ideas. My brain seemed full of building specifications and lime-mortar dust. I tried my best to think of something as I worked that day, but with little success. Then, just as I was getting ready to down tools, it hit me. Since I had little time for getting about, picking up on story ideas as I used to, there was only one thing to do: write about what was going on around me every day, the struggle to reclaim this place, the animals and my *Carry on Farming* approach to it all. I was sure it would work. Excitedly, I made a mental note to call some of the editors I worked with.

One was very interested. Another obviously thought that I was completely mad and kept asking me how on earth I was going to manage without a man – a pretty depressing response from a declared feminist.

Now I had to fit writing time into the days. Every morning I would crawl out of bed two hours before normal, often when it was still dark. Armed with the essential pint-mug of coffee, I would stagger blearily to my desk where I'd write until it was time to go building. In the evening, after the usual ritual of scraping off the mud, feeding the animals and having something to eat, I attempted to correct and improve what I had written that morning, often falling asleep with my head on my word-processor. I could only hope that this was the result of exhaustion and not boredom with my own words. I was so tired that it was getting increasingly difficult to be objective about my work.

When I first returned to writing, I felt restless, as if this sedentary process were alien instead of something I'd been doing more or less daily for the last twenty years. But eventually I began to enjoy the familiar routine of shaping and moulding a story. It was a little odd that it was my own story. I had never written anything so personal before. Mostly I had produced features, stories about someone else, other people's lives. Now the shoe was on the other foot and the feeling took some getting used to.

Despite my doubts, I had the first draft of a story written. It was much longer than the pieces I usually produced, but I discovered that I had a lot to say about Lickeen, my hopes and dreams for the place, the animals, my family, everything. Once I got the bit between my teeth I couldn't stop.

Sometimes, on those days when nothing went right and it seemed impossible that Lickeen would ever be fit to live in, I wished that all this were happening to someone else. But that mood never lasted long.

CHAPTER TEN

TOWN AND COUNTRY

The sun was out. It was hot, and there were even rumours of a heatwave, due to descend on us any day now from Spain, bringing with it fierce, hot air that would soon have us longing for rain again. I wore a dress for the first time that year, and felt strangely naked without the accustomed layers and wellies when I went into Bantry to do the weekly shop.

I had left Ray perched at a death-defying angle near the new chimney. He wore a battered old straw hat to protect him from the sun and he whistled cheerfully as he replaced the last few slates. Soon we would have a roof again and the thought cheered me no end. It had become difficult to imagine living in a house where water cascaded freely down the stairs when it rained, and whose kitchen was slowly developing a tidemark.

Bantry was crowded – farmers, housewives, holiday-makers and a gaggle of alternative lifestylers who congregated in the small square, playing penny whistles and juggling brightly coloured balls. Their dreadlocked hair and swirling

Indian prints added a certain exoticism to the monthly Fair Day. There were stalls selling second-hand tools, hens, cheap earrings, the contents of houses from people who had obviously decided to uproot and move on. Several sheep huddled miserably together next to a small cardboard box containing six bright-eyed collie pups. The box had 'Free to GOOD Homes' scrawled on it in a child's handwriting.

I wandered aimlessly for a while, looking around and thinking how strange it felt to be surrounded by so much noise and confusion, by so many people. The small town of Bantry had become the city to me and I wondered exactly when, on the long road from Los Angeles to Lickeen, this had happened.

Had I *ever* shot on and off freeways in Los Angeles? Or jostled my way amongst the packed throngs on Venice Boardwalk? My pre-Lickeen past had assumed a dreamlike and mythical quality. I longed suddenly for Lickeen's wide-open spaces and had to force myself to concentrate on the shopping list. Animal feed, fly repellent and worming powder came first. As I made my way through the narrow, crowded streets, I realized that I had come to hate shopping. It seemed such a waste of time and even more exhausting than knocking off rendering.

By the time I had got through the list, my small, battered Renault was groaning under its burden. The back was full of shopping and next to me, wedged into the passenger seat, was the yellow cylinder of gas that I had to replace weekly because there were no mains services in the glen. I had to get a prescription filled for a neighbour and then I could head for home.

Three women were in deep conversation in the chemist's shop, talking about the unexpected good weather, and wondering if it had anything to do with the Greenhouse Effect. 'Anyway girls, there's only one thing to do about it,' the eldest of the three was saying as she tugged at the hem of her multi-coloured floral print dress. The other two exchanged knowing glances as they waited for her to continue. The older woman gazed down with some satisfaction at her thick bare legs and heavily veined feet encased in slimline, strappy sandals. The white-coated assistant leaned across the counter so she could hear better. 'The only thing to do about this Greenhouse Effect,' the woman went on, looking from face to face to see that she had their full attention, 'is to make a good Act of Confession and enjoy the sunshine.'

I could still hear their laughter after I closed the chemist's door behind me.

When I got back to Lickeen, everything was quiet. Ray had left for his lunch and I took my own sandwiches round to the front of the house where the old stone would be warm against my back and the view across the mountains uninter-rupted. Everywhere I looked now there was extravagant, luxuriant growth and rich, deep colour. Wild fuchsia bushes clustered thickly around the house, their pink, purple and cerise flowers vivid against shiny green leaves. Eventually, they would have to be cut back because their riotous growth blocked much of the light from the small front windows. But for now I was content simply to enjoy them.

Looking at the glen from this vantage point was like contemplating some fantastic tapestry woven from many shades of mauve and green, with bright yellow gorse blossom running through it like exotic golden thread. I noticed that the heather was just starting to come into bloom again. When it did, I promised myself, I would go out onto the mountain and pick great armfuls of it, fill all my baskets and jars with its fragrant purple blossom which would still be vibrantly coloured by Christmas. Everything was so still and peaceful, and the sun felt wonderful on my skin.

I must have dozed off because I was suddenly aware of Kitty's rubbery lips brushing against my palm as she tried gently to lift the remains of a sandwich from me.

'You old rogue,' I said with no real conviction, realizing

that it was time to change out of my dress and get back to the real world.

Every morning now, my first stop was at Kitty and Anlon's field where I was sure I was going to find the new foal like the last time. Sometimes I took Kitty up to Lickeen with me, to Anlon's disgust. According to my calculations, Kitty had been pregnant for just over eleven months. When I discussed this with people who were experienced with horses they either said, 'She's too old, she'll never foal again,' or 'Oh, eleven months is nothing. I knew a mare who went for thirteen. You never know with a mare.'

They were right about that anyway. My farmer friends Dermot and Helen owned a mare who had been to a stallion and then scanned to see if she were in foal. The test proved negative. Nine months later, she produced a healthy filly.

I happened to mention Kitty's pregnancy to Finbar the vet one day when I met him on the forest road. His battered old car was piled up with syringes, assorted medicines and huge, calliper-looking instruments that I wasn't sure I wanted to know the purpose of. The inevitable cigarette was in his mouth, and he looked tired and rumpled. He told me he'd been up all night with a difficult calving. This was a busy time of year. 'That mare's not in foal at all,' he said, as he sorted through a tray of bottles for some wound powder I had asked him for.

In fairness, he had been of this opinion all along. But I couldn't help thinking of Anlon, Kitty's first foal, and how many people had thought the same thing then. This time I was so sure. A couple of local men who were working in the forest visited Kitty every morning, hoping to be the first to discover the new arrival. One of them swore to me that several times he had clearly seen the foal kicking inside Kitty. I told the vet this, anxious that he should believe it too, as if that would make this long-awaited event more likely to happen.

Finbar looked at me for a few moments, then suddenly he grinned. 'Yes, and there's a statue at Knock that'll move too if you stand looking at it long enough.' And with a wave of his hand and a roar of blue smoke from the exhaust of his station-wagon, he was off to his next port of call.

There were a few more weeks to go before I would give up hope. And then there was the question of Kitty's bulging udder, which, as I now knew, was full of milk. If it were a phantom pregnancy she was undergoing, then she was doing it in great style. The only thing I could do now was to watch and wait.

Since the weather had improved so dramatically the ground had dried out at last, and work on the boreen had been able to continue. But it was obvious that it was going to be much more than the two- or three-day job the contractor had originally estimated. Only the other day I had watched in horror as several lorry-loads of rocks, at seventy pounds a load, had been dumped into the mysterious holes in an attempt to fill them. The rocks had sunk without a trace, so the digger driver had had to take them all out again, excavate the holes even deeper to get rid of the soft clay, then dump them back in.

Springs had appeared where they had never run before, bogs opened up for no apparent reason, and no one was quite sure how the old stone bridge would stand up to being driven over repeatedly, if that day ever came. I didn't want to think of what the final cost of all this was likely to be.

Ray and I were talking about the continuing saga of the boreen as we were working one day when he suddenly said, 'Look at that wall, about halfway up. I think there was a little window there once. You can still see a small stone lintel.'

I looked over to where he was pointing and he was right. The lintel was about two foot in length and beneath it the stonework had been so neatly replaced that it would have taken his builder's eye to detect it. Since this was the outside wall of my soon-to-be bedroom, I was particularly keen to see what it looked like. With my new expertise at demolition, it didn't take long to knock the stones out and reveal the small window. It was about two foot square and, like the rest of the upstairs windows, almost at floor level. If you lay on your stomach there were great views of the cliff which towered above the house. Someone I knew in the village made beautiful objects from stained glass. I decided I would ask her to make me a pane for this little window.

Later that afternoon I went down to the car to see if the postman had left any mail. There were bills. Of course there were bills. I had almost become immune to their impact. And there was a letter from Amber, always a joy but today it was a particular pleasure to see her clear, distinctive handwriting in the middle of those unfriendly brown envelopes.

I opened her letter eagerly, hoping she was writing to confirm their arrival date.

Dear Mum

Hope you're not working too hard, although I expect you probably are. This is just a short note to let you know that help is on the way. We will be there on 3 September and this time you don't even have to pick us up because we are coming on the ferry in our NEW car. Kristine will fly from LA to Heathrow and we will collect her on the way.

I smiled to myself. I knew how much this old car – their first – and recently acquired from Bryn's uncle, meant to them.

We'll soon be roaring up the boreen ready to get to work, so just hang on.

Just then, I couldn't imagine them or anyone else roaring or even crawling up the accursed boreen, but there were still five days to go before they arrived and, as I had come to realize, a lot could happen in a short time at Lickeen.

I spent the rest of that day removing the rubble that had been deposited on the upstairs floors when Ray dismantled the roof. It was almost impossible to move around because of it, but Ray assured me that the lime-mortar would be great for firming up the boreen when the contractor had finished levelling it. So I dutifully hauled it down the narrow stairs in buckets and piled it up outside, trying to ignore the rising clouds of dust that had me constantly going to the bucket for drinks of water.

I worked late that night, possessed by the manic energy that occasionally overcomes me and usually leaves me incapable of any useful work for at least a week. By the time I got back to Barley Lake Cottage, it was nearly dark and my back and shoulders were aching unbearably. There was only one thing I wanted and that was a long hot bath.

CHAPTER ELEVEN

TALE OF A TUB

It had been three years since I'd had a bath of my own. No matter how I managed it, I vowed as I rubbed my aching neck, Lickeen would have a bath. I had got used to the outside toilet at Barley Lake, but I had never stopped longing for a bath. During this long, bathless spell, I would sometimes arrive on the doorstep of better-equipped friends, pathetically clutching my wash-bag and towel. It wasn't the same, though. In your own home you could make a night of it with a glass of wine and a good book, preferably when the wind was howling outside.

Not that it wasn't possible to keep perfectly clean without a bath. In the summer, after making sure there were no farmers with binoculars scanning the mountains for wayward sheep, I would duck behind the old stone wall outside the cottage and sluice off. In winter, I stoked up the old wood burner until it sizzled and, hoping there would be no unexpected visitors to my one-roomed cottage, I would strip off and wash in front of the fire. But it wasn't the same as total immersion in hot, or even lukewarm water, and a couple of times this craving landed me in some pretty tight corners.

There was the baby-bath incident for example. I found it on my weekly expedition to Bantry and for a baby-bath it was fairly large. I am not small, but I thought that with some judicious juggling of limbs it might just work. That night I boiled several kettles of water and got the wood-burner blazing. As I poured hot water into the turquoise plastic tub, steam rose up in satisfying spirals. Doing my best to ignore the cute, brightly painted ducks dressed in rainwear which adorned its sides, I carefully stepped in.

Try as I might, I could not seem to get my whole body immersed at the same time. If I managed to get my shoulders under the water, my feet shot out the other end. I made a last, desperate attempt. Bracing my feet firmly against the far side of the tub, I pushed down forcefully with my shoulders. The tub expanded obligingly and for one, glorious moment I felt hot water all over my body. Then the little plastic tub suddenly snapped shut again, leaving me trapped in a vice-like grip. Frantic contortions replaced those seconds of bliss until finally I lay, cold and exhausted, in a heap on the kitchen floor.

You'd have thought I would have learned from that experience. But I didn't. When I discovered an old oak rain barrel at the bottom of the field, I was sure I had found West Cork's answer to the California Hot Tub. I rolled the heavy barrel up the hill, deposited it outside the door and began clearing out the assorted wildlife, most of it slimy, that had accumulated there. Then I began boiling water. It was a very large tub. Six hours later I was still boiling water and sweating profusely as my small kitchen filled with steam. But as the long, slow dusk turned to blue-tinged darkness, I was finally ready.

Feeling more than a little ridiculous, I climbed up naked onto the kitchen chair I'd taken outside, put one foot into the steaming water then flailed around desperately. The wretched tub didn't seem to have a bottom. A black hole had obviously opened up in its murky depths. Then it began to lurch ominously. I wobbled with it and as I tried frantically to regain my balance, I remembered the vacuum cleaner incident.

I'd been cleaning my car. My neighbour down the hill had been in peaceful pursuit of his farmyard chores when he looked up to see a large orange vacuum cleaner hurtling towards him. I had made the mistake of letting go of it for a moment. Nothing was safe if left unattended on the short steep hill. The barrel lurched sickeningly for a second time and it looked as if history was about to repeat itself only this time it would be me, stark naked and encased in what would be left of the barrel, rolling to a halt outside my long-suffering neighbour's door.

It stopped rocking for a second – all the time I needed to retrieve the leg that was still dangling uselessly inside the tub, grab my towel and beat a hasty retreat for the great indoors, to await better days.

CHAPTER TWELVE

HOME FOR THE HOLIDAYS

It had come to the point where there was no more that the man with the digger could do for my boreen. Thanks to the continuing good weather, he had filled in holes, moved boulders and levelled out the track's more challenging gradients. Now it was up to Ray, my friend of many talents, to make it driveable. He had to spread trailer loads of gravel, fill in ruts and create a surface that car tyres could get a grip on. When I looked at its scarred and pitted surface, I couldn't for the life of me see how he was going to do it.

Damn the boreen anyway. By the time I got the final bills for all this work, I was going to have to go back to the bank manager again for help. There wasn't much else I could do. Until it was driveable, the essential services I needed – telephone and electricity – could not be connected. You could see the workmen's point. I wouldn't much fancy trudging up the boreen with huge, cable-bearing poles either.

I sat slumped over my tea at Barley Lake Cottage surrounded by bills. I couldn't continue in this hand-to-mouth fashion. After we had bought Lickeen I had naïvely imagined that the biggest struggle would be renovating the place. Now it looked as though the real fight was going to be trying to hold on to it. I added, 'Call Bank Manager' to tomorrow's list. Even now, I couldn't completely abandon myself to despair, or contemplate selling up and getting out while I was still sane. The stubborn thought remained that if I did right by Lickeen, then it would do right by me.

'Fine evening.' Ray's head poked round the open front door.

I agreed, automatically reaching for the teapot to pour him a cup.

'No, thanks, I won't.' He raised a declining hand. 'I only called in to tell you that you can drive up your boreen now – if you want to, of course,' he concluded nonchalantly.

If I wanted to? I had been dreaming about it for weeks. I had never imagined he would be able to pull this off so fast. If I was honest, I hadn't thought he would be able to pull it off at all. When I had left Lickeen that afternoon, Ray had still been spreading gravel.

I had to try it out – right now. I wanted to drive all the way up to my own front door. Ray understood when I hurriedly excused myself and ordered the dogs into the back of the car. They would have preferred the boreen to have remained undriveable for ever, of course. Tom and Sam have never been able to see the sense of driving when you could be walking, running, sniffing. But they jumped in anyway, pained expressions on their usually good-natured faces, and we set off.

Other than a bit of wheel-spinning on the freshly laid gravel, I was indeed able to drive straight up to the front door. I sat in the car, beaming idiotically. Then I repeated the whole performance all over again. I just couldn't believe it.

We had driven up and down several times in rapid succession before I had convinced myself that it was really true. The dogs' expressions of resigned acceptance had changed into ones of sheer anguish. They obviously thought I was going to be at this all night and they wanted out. Reluctantly I turned the car towards one of our favourite forest walks, amazed at the thought that tomorrow I would actually drive to work.

When I arrived at the entrance to Lickeen the next morning, Ray was already there, in deep conversation with the lorry driver from the builder's suppliers. He had another load of materials for us: Ray was going to be building a septic tank next, and the lorry groaned under the weight of hundreds of blocks. Ray was telling the driver that he could take the supplies all the way up to the house now. From the look on his face I could see that the driver was far from convinced. The state of my boreen had become legendary. During one particularly heavy rainfall, a sizeable portion of it had washed down into the road where passing traffic had had to make a detour in order to pass. 'It's fine now, grand,' he was saying to the driver, who was looking uncertainly up the quarter-mile stretch. 'Why, 'tis better than the M1!'

The driver grinned at him. 'I'll chance it so,' he said, and with a grinding of gears, he nosed the huge lorry cautiously into the gateway and slowly up the boreen. I followed behind at a respectful distance. I don't know about him, but I was holding my breath, especially when he inched his way carefully over the old stone bridge. When he made it, all the way to the top, I felt like cheering. As far as I was concerned Ray was right. This boreen was better than the M1. I decided I wouldn't tell the others the good news. They were due to arrive in two days anyway. I would surprise them by driving us all up to the front door.

On the day they arrived, the dogs were already down at the bottom of the steep hill outside Barley Lake Cottage, barking furiously, before I realized a car was coming. I'd overslept and had spent the morning rushing around cleaning, baking, and seeing there was enough bed-linen for everyone. Now the cottage was gleaming, and smelt of fresh bread.

I hurried to the front gate just in time to see Amber, Bryn and Kristine chugging up the hill in their little blue car, waving at me and calling the dogs who were beside themselves with joy.

We hugged and talked, all at once, and then we went inside for tea. Later I took them down to the neighbour's field where Elly was grazing, for formal introductions. It was as good a time as any, I decided, to break the news of Elly's unexpected pregnancy. Knowing how Amber worries that I'm constantly getting myself way out of my depth, I had put off telling her of this startling development by letter.

Everybody ooh'd and aah'd over Elly's sleek red coat and long curling eyelashes. Elly preened herself under this onslaught of attention.

'Don't you think she looks a little on the heavy side?' I asked, as casually as I could. Everyone looked at me curiously.

'The thing is, she's in calf,' I blurted out.

'Isn't that going to be a lot of extra work for you?' Amber asked, eyeing me speculatively.

'Well, not really,' I protested. 'And I'll probably sell the calf when it's about six weeks, so it will bring in a bit of extra money.'

One look at Amber's face convinced me that she seriously doubted my ability to sell any animal once I had fallen under its spell, as I had with this little cow.

I was obviously going to have my work cut out convincing her that, despite my past track-record, I was determined to be a responsible and profit-conscious farmer from now on. The fact that Elly was engaged in affectionately licking my knee with a tongue rough enough to lift tar off roads probably didn't help my case either.

Amber grinned and linked her arm through mine.

'Oh, Mum,' she said, shaking her head.

*

I sensed a subtle difference in the house as soon as I awoke the next morning. Since Amber had begun college several years before, I had become used to living alone. Most of the time I wallowed in the feeling of being able to please myself about when and what I ate, where I went and when I came back. But there were still times when I felt a real sense of loss that Amber was no longer a daily part of my life. Now I felt a rush of pleasure as I remembered that she was asleep in the next bedroom, only a few feet away.

I crept quietly downstairs, closing the trapdoor to the attic bedrooms carefully behind me. I had decided to make some muffins as a treat for breakfast. Soon the warming scent of carrots and raisins filled the kitchen. It seemed strange not to be getting sandwiches and supplies ready for the day's work, but the others wanted some time to unwind from city stress, to slip back into the gentler rhythms of West Cork, so I was having a break too.

I smiled to myself when I remembered the looks on their faces when we had gone to Lickeen after visiting Elly. Instead of parking and walking as they were used to, I had opened the gate, got back into the car and kept driving, all the way up to the house. Stripped of its rendering and partially pointed, the house was a surprise to them too. There were six new windows, new doors, a neatly retiled roof, a chimney stack that was whole, and Ray was almost halfway through rebuilding the porch. Much of the dank growth that had surrounded the place was gone now, and the building stood out, dignified and imposing against the mountains. They were ecstatic, amazed at how much had been accomplished since they had last seen it. Looking at it through their eyes helped me regain my own sense of persepctive.

We took it easy over the next few days, talked a lot and caught up on everybody's news. Without wishing to worry them too much, I told them that I had been to the bank manager and, to my great relief, he had agreed to bail me out once more.

I looked across at Amber's concerned face. We were all sitting round the kitchen table, drinking coffee and chatting. Amber's dark curly hair was pulled back from her face in a pony-tail and her deep brown eyes filled with concern. We had survived some pretty tough times together and enjoyed some quite spectacular ones too. Throughout them all, this poised young adult before me had always shown a touching confidence in my ability to survive. Sometimes I felt it was misplaced. I had not married Amber's father and, having decided to go it alone, had spent many hours anguishing over my ability to raise her. As it turned out, over the years we brought each other up. The bond between us was deep and very precious.

When I had finished explaining my current financial situation, everyone was quiet for a few moments. Then Amber smiled and said, 'Let's talk about work now. What shall we do first?' I leaned across the table and gave her a quick hug.

We were a motley but cheerful crew the next morning as we set off for Lickeen, everyone wearing borrowed garments from my plentiful supply of work clothes. We had sand and cement, and a mixer that had to be started with one of those cursed pieces of string. Together we were going to tackle the pointing.

Soon everyone had found their preferred tools for the painstaking job of cleaning out the cracks between the stones before packing them carefully with cement. Spoon handles, chisels, kitchen knives and, inexplicably, an egg-slicer, were just some of the instruments we worked our way through. When it came to filling these cleaned-out gaps, though, there was little argument. Everyone agreed that the human hand was the most efficient tool, a decision I'd reached after pointing for only ten minutes.

The technique I had evolved was basic but effective. Hefting your bucket you stationed yourself in front of your newly cleared patch. Making sure that you were wearing a

stout rubber glove you picked up a good handful of cement
and sort of mushed it into the gap, pushing it as far back as
you could with the heel of your hand. Later you went over it
with a damp brush to smooth it out. We made great progress
that day.

Amber, Bryn and Kristine took to the navvying life as
though born to it and went off to Lickeen each morning
leaving me behind. I was writing in the mornings, building in
the afternoons. The first story I had completed on Lickeen
had been well received by the newspaper. They had requested
photos and a follow-up.

Even though I had appeared for the photo session at
Lickeen hot, flustered and late, wearing a pair of jeans that
are almost impossible to sit down in and therefore nearly
always clean, the pictures came out well. Not that I had any
doubts about my photographer colleague's professional capa-
bilities. It's just that to my eye, at least, I usually look slightly
demented in photos. This time, there was a mild, tranquil
sort of expression on my face that I didn't instantly recognize
when I looked over the prints. It seemed to have little to do
with the last crazed few weeks. The dogs lolled about looking
soppy and appealing and Kitty stared into the camera while
chewing thoughtfully. We all looked as if butter wouldn't
melt in our mouths.

The two weeks we were all together sped by. We had
walked the land at Lickeen, planning what we would do if
only we had the money. We pulled some of the briar and
furze off the old ruin above the house that would someday
be Amber and Bryn's home, and found a tiny and beautiful
blue glass bottle amongst the crumbling stones. We had
visited the site where Kristine wanted to build, and had
marvelled at the two-hundred-foot sheer cliff-face which
formed part of her boundary. Kristine lived too far away to
be able to visit as often as she would like; Amber and Bryn
had college and the challenge of trying to make ends meet on
student grants. It was the first chance we'd all had to get

together at Lickeen since we had bought it, and we made the most of it.

Between them, they fed the cow, saw to the horses and frequently cooked delicious meals. I was going to miss them. Most of all I was going to miss the wondrous talks we had, gathered around the kitchen table, dishes unheeded in the sink, wide-ranging, invigorating talk that often went on late into the night. It wasn't the time for them to make a full-time commitment to Lickeen, they were all too involved with sorting out their own lives and shaping their futures. But I was determined Lickeen would always be there for them, for all of us.

Even the dogs looked dejected on the day they left, lying quietly next to each other by the wood-burner, pointedly ignoring the luggage that was now heaped in the middle of the floor. It was always the same, I thought glumly, as I trailed behind them to the car carrying someone's backpack. I'd promise myself I wouldn't get emotional and then without fail, I did. But I wasn't alone. We all gulped and hugged and cried a bit, then hugged some more. Eventually the blue Datsun Cherry roared off down the short steep hill from Barley Lake Cottage. I tried to cheer myself by thinking that we were planning to get together again at Christmas. It wasn't all that far away.

The dogs thumped their tails feebly when I went back indoors. They were obviously depressed too. The cottage had that empty echoing feeling that you get sometimes when you are moving, the hollow sound of newly created space. We could all do with a walk I decided, giving up on my half-hearted attempt at tidying away the dishes.

At first the dogs were less than enthusiastic. Like me they would have preferred to mope around the house for the rest of the day in front of the wood-burner. But we all brightened up a little after we had gone a couple of miles up the Barley Lake Road. A stiff breeze was blowing off the Caha mountains and the sky hung dark and low over their jagged cliffs.

The wildness of this part of the glen was both sombre and exhilarating, especially on an emotional day like this.

I sat and stared at the grey ruffled waters of Barley Lake, my mind a blank, until the first heavy drops of rain drove me towards home and the warmth of the fire. Tomorrow, the plumber was coming and it would be business as usual. I wanted to be ready.

Chapter Thirteen

SUMMER'S END

September was drawing to a close, and still Kitty hadn't foaled. It was time to resign myself to the fact that she was not going to. This was a classic case of phantom pregnancy complete with a plentiful supply of milk. When I met the blacksmith in the village one day, he told me that he'd heard of a number of mares who had gone through this strange and mysterious process lately.

It was a big disappointment, particularly since at Kitty's advanced time of life there was no guarantee that she would foal again. Anlon might be her only offspring, a wonder colt born when nobody had expected him. The breeding season was over now so I could only wait until next spring to try again. In the unlikely event that a foal were conceived this late in the year it would be born too late to benefit from that vital flush of rich grass in spring and early summer.

I sighed and rubbed Kitty's strong neck. She nibbled enquiringly at my pockets. Despite the work involved in raising a foal, I had been looking forward to having one gambolling about the place again. But when it came to animals, as I was learning daily, there were no sure bets. I may have been farming on a very small scale but every day I was having to learn the patient acceptance which was such an important part of the whole process.

There was a bonus, though, to this phantom pregnancy that I had overlooked. Now I would be able to go riding again. Owing to her 'delicate condition', Kitty had been neither worked nor ridden for several months, though she had been receiving special rations which she had, of course, wolfed down with considerable relish, sniggering behind her hoof, no doubt. Since she wasn't being ridden, her shoes had been removed some time ago. She would need new ones before I took her out on the roads again.

Having the blacksmith come to call was always a special occasion for me. He was the last man around these parts still performing the dying art of hot shoeing. Since he didn't drive, or have a phone, the first phase of the operation was tracking him down. This usually involved calling his local pub and leaving a message.

The first time I saw him at work I was more than a little confused. Calmly he produced a small Hibachi grill, a vacuum cleaner and a bag of coal. I thought for a moment we were about to have some sort of impromptu picnic. This equipment turned out to be an entirely ingenious solution to the problem of the vanishing forge, once the centre of village life.

His vacuum cleaner was of the old cylinder variety and there was an aperture for its nozzle in the back of the grill. After he had prepared firelighters and some small pieces of wood, he would fix the hose into the back of the grill, plug the vacuum cleaner into the nearest electricity outlet and create a fine draught which soon had the fire blazing. At just the right moment he would add some coal. Soon the fire would be hot enough for the skilled task of shaping the shoe exactly to fit the horse's hoof.

What this man didn't know about horses and their feet was definitely not worth bothering with. And the stories that came with the after-shoeing refreshment were magic – strange doings of horses in times long past, tales of the village when the forge was its meeting place and the blacksmith its king, and of his great-grandfather, renowned for crafting superb wrought-iron gates, one of them still standing today.

I telephoned the blacksmith's local pub and he returned my call almost immediately, saying he'd had a cancellation and would be free the next day if I could go and pick him up. Usually I would have had to wait for a couple of weeks as his services were much in demand.

Outside the front door of Barley Lake Cottage the next afternoon, Kitty amiably leaned her not inconsiderable weight against him as he worked on her large feet. Sweat ran freely down his face and the cold air was heavy with the smell of burning hoof and chilling metal. When he had a shoe shaped to his satisfaction he would plunge it into a bucket of cold water. After that came the tricky business of nailing on the shoe. One slip could send a nail into the 'quick' of the hoof and leave the horse permanently lamed.

The smoke from the blazing coal fire blew into our eyes and noses as he worked on. The dogs mooched around our feet searching for bits of Kitty's hoof that the blacksmith had previously pared off. They found these morsels irresistible. When I first saw them munching away enthusiastically it seemed a bit like someone gobbling the bits after you've cut your toe-nails – slightly sickening. But the blacksmith explained to me that they were very high in protein and most dogs, quite sensibly, loved them.

Kitty looked resplendent in her four shiny new shoes, hoofs neatly trimmed and pared, ready to tap-dance her way down many a path and trail with me. After the tea and stories we loaded the blacksmith's gear into my car and drove back to his village where he had another customer due at any moment. I was anxious to get back home myself and ride Kitty along one of the forest paths before it got too dark.

I was in the process of looking for her tack, which had not been used in quite a while, when the phone rang. At first

I ignored it. Then I thought perhaps it was a newspaper editor wanting to offer me a well-paid, stimulating assignment. Stranger things have happened.

I didn't recognize the voice on the other end of the phone but I could tell it was an elderly man. He said he was calling from Limerick. He had read the story on Lickeen in the paper and wanted to tell me how much he had enjoyed it. I thanked him. 'Anyway, the thing of it is,' he continued in a wavering voice, 'I've got quite a bit put by, and I'm a great man for the cattle, you know.'

I murmured a polite response, something along the lines of being quite fond of them myself, not quite sure where this conversation was leading and with an anxious eye on the darkening sky.

'Well, I'll get to the point,' he continued briskly. 'It looks to me as if you could do with a man about the place, and I was thinking I might pay you a visit. There's just one thing though.' He paused for a moment obviously having a bit of difficulty with whatever it was he wanted to say next. 'I was just wondering,' he eventually blurted out, 'would you still be of child-bearing age at all?'

That he was probably well into his eighties seemed to trouble him not at all, and, for all I knew, perhaps he was well able for any and all of what he was proposing, but for once I allowed discretion to be the better part of valour. I thanked him for his phone call and managed to put him off by indicating that were I not otherwise engaged I would have been charmed by his proposal.

While Kitty, the dogs and I were ambling through the twilit forest later, at peace with the world, I couldn't quite rid my mind of the nagging thought that one day, when I was least expecting it, I was going to look out of the window to see a lively octogenarian farmer striding determinedly up the boreen to claim my hand.

*

I was currently engaged in a project that these days struck me as almost lady-like – tiling the inside kitchen window sills with pieces of grey slate left over from the roof. It was easy work and very satisfying. The old kitchen had come a long way since those early rubble-infested days. Now there was plumbing, four radiators, a kitchen sink and, outside my new front door, a large, ancient cast-iron bath tub that until a couple of weeks ago had been used for watering cattle in a neighbour's field. I had bought it for twenty pounds and once it had been cleaned up it looked splendid. I couldn't wait to see it installed.

The outside pointing was finished and I had removed most of the plaster inside. Ray had built a septic tank at the bottom of the front garden, and next week he planned to start making a small bathroom upstairs. I was beginning to look forward to the fun stuff – painting the walls, arranging my belongings, and lounging in front of the huge fireplace with a large whiskey.

I stood outside looking over the mountains after I'd finished tiling. A sudden gust of wind deposited a shower of ginger leaves at my feet. The summer had sped by. Slowly, almost imperceptibly, gold, russet and amber tones had replaced the many shades of green. There was an unaccustomed sharpness in the air and the scent of damp leaves mixed with the heady aroma of overripe blackberries.

Last week, I had been in Glengarriff village for supplies and got stuck behind what were probably the last tour buses of the season, trying to perform complicated manoeuvres on the narrow main (and, indeed, only) street. Visitors in brightly coloured shiny tracksuits sipped pints outside the pubs, enjoying the warm autumn sunshine, and the small row of gift-shops at the far end of the village were still doing a roaring trade. Soon the tables and their umbrellas would be taken inside and most of the gift-shops would close. As I had waited to be served in the small village store, I wondered whether the other customers had been as surprised as I was to discover that they had been able to buy fine wines, pâtés, tofu and many other exotica in this out-of-the-way spot.

In the winter, when the proprietor knew everyone's face, the little shop also offered a good selection of videos. It was then that I caught up with any films I wanted to see, since a visit to the cinema involved a round trip of over a hundred miles. Before I had moved to Ireland, I had bought a small television with VCR combined, and watching videos in my little cottage was a winter treat I looked forward to. It would be especially satisfying in the front room at Lickeen with the fire blazing.

This seasonal flood of visitors to Glengarriff contributes more than much-needed income to the area. The visitors are a rich source of interest and diversion after the long slow winter months when nothing moves and you recognize every car you see driving along the deserted roads. One particularly colourful group descended on Glengarriff a couple of years ago and decided to stay awhile. They were Hare Krishnas who decided to buy a remote, mountainous farm not far from Lickeen. Locals watched them settling in with undisguised glee. At night the pubs were full of stories about their chanting, vegetarianism and strange practices, but, eventually, the sight of them trudging down the rough boreen headed for the village with their saffron robes tucked neatly into their wellies became an everyday occurrence. People even came to accept that after one of their mass head-shavings you wouldn't be able to buy a razor-blade in the village for weeks.

One day a young Krishna devotee was on his way to Glengarriff when he encountered a local man. The boreen was narrow and he stepped aside to let the older man pass murmuring, 'Hare Krishna,' in customary greeting. Determined not to be outdone in courtesy by a stranger, particularly one wearing an orange robe and green wellies, the local man responded gravely, 'Harry O'Sullivan, 'tis a fine morning,' and went on his way.

Another story, still fondly retold, concerns the time the

Krishnas bought a horse from some passing tinkers. The sturdy cob was with them for less than twenty-four hours before, with the unerring instinct of a homing pigeon, he returned to his original owners, something he had obviously done many times before. Needless to say, the tinkers didn't hang around to argue the point.

The Krishnas eventually moved on to an island off the North Coast, so it was said, where they were able to reactivate their British tax-exempt status. But their memory lingers on.

I sat outside my new front door for a while, watching the sun dip behind the mountains, holding my breath for the last vanishing rays which would transform everything to molten gold. For a glorious moment I was filled with longing to be living here, to feel the silences last thing at night and see the majestic views first thing every morning. If everything went according to plan, I wouldn't have to wait much longer. I hoped to move in the week before Christmas.

I walked round to the barn at the back of the house where Kitty was steadily munching. It was time to go back to Barley Lake. Attaching the lead rope to her head-collar, I whistled for the dogs and started down the darkening boreen.

I had cleared out most of the debris inside the barn – rocks, trees and assorted bits of ancient farm machinery. Now it was quite spacious, perfectly big enough for my ménagerie. It would be the ideal spot to store the hay that I would soon be needing. The barn was a low, single-storey structure built of the same stone as the house. The whole thing had been erected on a huge rock which formed the main part of its foundations and which sloped steeply down from the barn doorway towards the house. Outside, several large cup-shaped stones protruded from the wall, forming natural seats that looked as if they had been worn smooth by generations of backsides. Since Kitty and I had cleared much of the undergrowth that had almost covered it, I enjoyed sitting outside the barn, imagining the many animals, particularly horses, that must have sheltered under its roof over the years. I had found several horse-shoes of all sizes inside.

I had wanted to find some old slate for the barn roof. But, realizing that I wasn't going to be able to afford it, had had to resign myself to aluminium sheets, determined that after they had weathered sufficiently I would paint their garish surfaces some sympathetic colour. The animals wouldn't care much anyway. The old barn was going to make fine, snug winter quarters for them.

It was only about fifteen feet from the house and Ray was always saying that he thought this was too close, but I didn't mind at all. I already knew that, years ago when such things had been acceptable, normal even, I would have been the kind of farmer that not only had the barn near the house but who frequently welcomed its occupants inside. Besides, with Elly's unpredictable calving expected early next year, there would be distinct advantages to having this building so close to my back door.

When I moved into the house, I would bring the animals up to Lickeen. What was going to happen when Kitty and Elly, now a very jealous and possessive little cow, finally came face to face, I couldn't imagine.

CHAPTER FOURTEEN

BARLEY LAKE

Plastering defeated me. I had managed not to say, 'You must be joking' when Ray had shown me the newly erected scaffolding I had to scale to knock the rendering off the top of the house; I had learned how to pick up unwieldy bags of cement, and knew what was the best mix for pointing walls. I had even learned how to start the cement mixer – at a pinch. But, try as I might, I could not master plastering. Whenever I threw a lump of the thick grey mix onto the wall with what I hoped was a plasterly flick of the wrist, it hung poised there for a few moments then slid slowly off again, hitting the ground with a depressing splat.

The few stretches of wall where I had managed to make some of the stuff stick looked like the surface of a cake that has been got at by an insane icer. It didn't help that they were next to the smooth and apparently effortless stretches that had been glided over by Ray. I hated everything about plastering – the mixing, the smell, the mess – so I avoided most of the inside walls. Removing the stuff from them had been quite enough.

Then one day in late October, massive poles of the cable-bearing variety mysteriously appeared on the forest road. Electricity was at last on its way to Lickeen.

I confidently expected there to be crews of eager workers following close behind. Excitedly I called the electrician and asked him if he would come and do the wiring right away. He did, but several weeks went by and nothing else happened. The poles just lay there apparently forgotten. I made several frantic phone calls to the power company. They promised that it would be any day now.

Meanwhile our working lives were becoming increasingly difficult. Sometimes, even at midday, it was impossible to see what you were doing in the dim, misty autumn light. Often I had to hold a torch while Ray tackled a particularly dark corner. We had intended to plaster after the electrician had been but Ray had to go to another job for two weeks and I needed to press on so we had decided to do the plastering before he left and have the electrician lay his wires over it. This seemed a perfectly reasonable compromise to me and the wires could always be painted over.

But the electrician had refused to adopt this sloppy technique and painstakingly chipped off strips of newly laid plaster in which to set his wires. Now somebody had to cover them up again before he could finish the job and I could get a certificate of safety that would finally enable me to get an electricity supply.

There was nothing else for it. Since Ray wasn't around I would have to do it. The light inside the house was appallingly bad that day and it was, of course, raining, so I had to make up a batch of cement on the porch. I scrabbled around in what was left of the lime and threw several handfuls in a rather haphazard fashion into the sand and cement. This is the same system I apply to baking, and more often than not it works pretty well. But my flour rarely has wood-chips or small pieces of rendering in it. I thought the mixture felt very grainy as I was ineptly trying to apply it to the wall, but by this time it was virtually impossible to see anything so I contented myself with the thought that I could always brush off the worst of the lumps and bumps with a yardbroom the next day.

In the cold bright light of the following morning I had once again to face the results of my handiwork. I spent several hours picking woodchips (which the lime had been full of) out of the now firmly set cement. They looked like confetti and gave the house a curiously festive air. But there was nothing I could do about the hundreds of tiny pieces of rendering now firmly stuck to the walls. They would be there to haunt me for ever, and no amount of smoothing was going to remove them.

In the midst of the mayhem that was still Lickeen, I would sometimes stare out over the glen transfixed. It glowed with a last brave burst of colour. I thought about the year that was drawing to an end. It had been a strange and wonderful time, exhausting and exhilarating. Soon I would be living here, passing my days deep in the mountains, lulled to sleep during long winter nights by the sound of the wind sweeping down from the Caha mountains. I couldn't wait.

But before that day finally came, I was feeling in need of a break. Between working on the house, caring for the animals, and writing the stories now winging their way to various editors' desks, I was feeling a little frayed round the edges. Work had finally begun on erecting the electricity poles, cables and transformers needed for my supply, but that was turning out to be a more difficult job than the electricity company had expected. I was not surprised. The ground where the poles were having to be placed was rocky and overgrown. Sometimes I was afraid the workmen were going to give up in disgust and walk off. If I stayed at Barley Lake Cottage, tried to lie around reading a book, say, I would only feel guilty about the many things I should have been doing. I had decided to have a proper day off and take the dogs and myself to Barley Lake.

Barley Lake is a remote, glacial, heron-haunted body of

water, in an austere and beautiful mountain setting, where I have dreamed away many happy hours, but also where, on odd occasions, I've become strangely uneasy and have been maddened by a sudden impulse to turn round and look over my shoulder, even though I knew full well that nothing was there. Some believed that the souls of the dead resided in lakes. And perhaps they did – there are many local stories about this. One of the best-known is the tale of Duffy the Boatman who has sometimes been seen waiting patiently in his small boat for passengers. Those who have been foolhardy enough to accept his silent invitation have never been seen again.

One evening when I was up there with the dogs sitting about enjoying the view, in that shape-changing twilight I saw what looked like a fleet of small boats moored at the far side of the lake, each with a large Cyclops-like eye, similar to boats I had once seen in Malta. I blinked hard and looked again. They were still there, bobbing gently on the waves. The dogs didn't seem to have noticed anything strange. They were sprawled out on the rocks, sleeping peacefully, and were obviously going to be no help at all in determining if we were in the presence of psychic phenomena. I forced myself to look again at the darkening lake. As I did, the boats were caught up in the dying rays of the sun which had turned the lake to flaming red. For a moment this fire seemed to lick hungrily at them – and then they were gone. There was only the dark, deep lake surrounded by empty, jagged cliffs just as there had always been, and the heron, balancing on one slender leg, waiting patiently in the shallows.

But psychic phenomena aside, Barley Lake has a special appeal that keeps pulling me back to its pebbly shores. Even in the summer it is a supremely peaceful place. Most visitors who negotiate the hairpin bends on the narrow mountain-road never get much further than the cliffs overlooking it. To reach the lake you must first cross a lengthy stretch of rough and boggy ground. I have seen visitors take one dubious look at this steep descent and another at their footwear and think better of it. So you can usually count on having the place more or less to yourself.

The dogs and I ignored the few dark clouds which scudded across the early-morning sky as we prepared to head off. I closed the cottage door firmly behind me and picked up my backpack.

We made our way up the narrow winding pass at a steady pace. Winter had already taken hold up here. The mountains looked threadbare and brown and had assumed the appearance of patient waiting that signals the onset of winter. Around the bogs, where several of my neighbours still cut their winter's fuel every year, lay scattered pieces of turf that were too small to be worth gathering. They were a dark wet brown. Ahead of us, the mountains seemed to stretch endlessly, silent and utterly still.

When we reached the top of the road the dogs and I rested amongst a tumble of rocks which overlooked the lake. They were bright with green and red lichen, swirling patterns which looked like elaborately applied silk-screening and felt surprisingly warm to the touch. If I looked to my left I could see the dark winter green of fir trees in Glengarriff Forest, and, beyond them, the deep curve of Bantry Bay.

I watched transfixed as its sheltered waters turned from steely-grey to a light, shimmering turquoise illuminated by the pale winter sun.

Satiated with silence I dropped down in my perch to observe the clouds high above me, my head resting on the book I had lugged up here, knowing that I probably wouldn't open it. Later, I planned to walk around the lake to a pocket-handkerchief sized beach where I liked to sit and throw stones into the water for the dogs. But for now I was content to do absolutely nothing. This would probably be my last idle day for some time and I was determined to make the most of it.

Once, the Lords Bantry owned all this untamed land as

well as Whiddy Island, Glengarriff Forest and a lot more besides. In those days the aristocracy didn't miss a trick. One of the Bantry clan built a small rough lodge, and a crowd of them would ride up on their fine horses and hunt hares all through these wild mountains. At night they would build huge fires and feast on the day's kill. Today the lodge is nothing more than a pile of stones, and I have never seen a hare up here.

Early one morning I did see two sleek young otters having a fine breakfast of fresh lake trout on the little pebbly beach. How the fish got here in the first place and how they manage to thrive not only in this lake, but in the three hundred and sixty-five others that stretch down the bony spine of the Beara Peninsula, I do not know. For, unlike many other local lakes, they have never been stocked.

I lay contentedly curled up in my cleft of rock puzzling over this mystery until sleep overtook me and I dreamed of great rainbow-coloured fish with one Cyclopean eye, ancient creatures equally at home on land or water who moved freely from lake to lake in shining migratory trails.

When I finally awoke I felt rested and totally refreshed, as if I had slept for several days instead of just over an hour. The dogs looked at me hopefully, wagging their tails. They wanted water and stones and long hikes over distant mountains. And just then, so did I. They spent much of that afternoon chasing a rabbit who was obviously only toying with them. When we got back to Barley Lake Cottage they immediately stretched out in front of the fire and fell sound asleep. Occasionally they twitched and growled, dreaming of the one that got away.

In preparation for my move to Lickeen, I decided that evening was a good opportunity for sorting through my boxes of photos. I'd been lugging them around for several years with every intention of transferring them into a photo album. For once I wasn't exhausted. I carried the two boxes downstairs and dragged them over to the hearthrug to join

Tom and Sam. At times like this, the dogs made pretty handy occasional tables. Soon they were completely covered in photos. Neither of them stirred.

Amongst this morass of images from over the years, I came across a batch of pictures taken in Los Angeles during the time I had worked for the *National Enquirer*. I looked at them curiously. They bore no resemblance to the images of me that had recently appeared in the paper. You wouldn't have known it was the same person.

One photo from the good old days that really made me cringe was of me dressed in a hideous pink exercise outfit, all ready to work out with Lou Farrigno, better known to his many fans as the Incredible Hulk. I remembered that unfortunate incident as if it were only yesterday.

Lou had opened a new gym in Santa Monica and was presumably hoping to convince the body-obsessed Los Angelinos that if they paid him vast sums of money they too could look like him with bulging biceps and a shirt that ripped open obligingly in times of stress. Actually, I remembered Lou as being a very nice guy, one of the few celebrities who didn't reach for the Holy Water when they heard the words *National Enquirer*.

When Lou's agent had contacted the paper about this story, they had assigned me, intrepid girl reporter, to cover the event. In the photo I was holding Lou is grinning widely, as well he might, for I am stretched out across one of his gleaming chrome torture machines looking for all the world like a very angry beached whale encased in pink lycra. A friend who was always dieting kept this picture stuck to her fridge door, claiming to find it a great source of inspiration when temptation struck.

It was hard to believe now that I had ever gone skulking behind palm trees in expensive Beverly Hills restaurants or eaten extravagant meals on *Enquirer* expenses in the vain hope that some drug-crazed celebrities would fling themselves down at my table and insist on telling me the sordid secrets of their sex lives. Nowadays the nearest Chinese takeaway involved a round trip of 120 miles. Eating out had come to mean a salmon sandwich and a pint at the pub. Looking at this picture I couldn't really say that I was sorry. Today I was a lean, mean fighting machine who had developed unexpected reserves of physical strength that would have impressed even old Lou, if only he could see me now.

By midnight I had sorted through several years' worth of photos, discarded hundreds and had one box neatly stacked in a corner. It struck me that this was it, the first of the packing. Sooner rather than later, I would be moving. I was going to miss Barley Lake Cottage. Slowly I put the last of the unwanted photos in the fire, watching them brown and curl at the edges before suddenly bursting into flame. It looked as though we were going to be spending Christmas at Lickeen after all.

CHAPTER FIFTEEN

AND THEN THERE *WAS* LIGHT

Elly, usually so nimble and sprightly, was becoming ponderous. Although it was only November she looked enormous. When she lay down in the field with a grateful groan she appeared to treble in size. She seemed much too big to last until March when the calf was allegedly due, and her udder had begun to distend and fill. A neighbour remarked that it looked like 'a bunch of baby carrots'. The general consensus was that she would make a fine milker. Whether I would make a fine milkmaid remained to be seen.

I couldn't rid myself of the images of calving ropes and jacks, implements that are sometimes needed in a difficult birth when the calf might be too big or turned the wrong way. Sometimes I would dream of twin bull calves that were as large as Elly at birth, with shoulders the size of American football players.

It seemed years ago since I had made that innocent application for electricity. There were times when this whole situation seemed at least as bizarre as a story you might find in the *National Enquirer*. I comforted myself with the thought that this was a time in my life for learning new and apparently unrelated skills.

Despite my anxieties, Elly thrived. Her sleek red coat thickened and curled and she gleamed with health. I noticed when she spent a few days at Barley Lake that she was eating more than enough for two, eagerly cropping down the grass that Kitty and Anlon hadn't eaten. Fortunately for me, horses and cattle are a good mix for grazing purposes. Each will eat where the other will not and so the end result is a well-manicured (and well-fertilized) lawn. They also absorb and destroy each other's parasites. When I first began reading up on animal husbandry, I was horrified to learn how many different and disgusting kinds of worm-infestation livestock are prone to. It's the sort of information that could keep you awake at nights, so any help in keeping the beasties down is a definite bonus. I kept a regular record of when I dosed the livestock, and entries such as 'Elly – liver-fluke' would appear regularly in my diary next to notes about story dead-lines.

When I had the time I would brush Elly with the stiff long-bristled dandy brush I used for the horses. She loved it, arching her back and headbutting me gently for more when I stopped. I had tried attaching a lead-rope to the head-collar she sometimes wore and walking her around the field, a handier approach than trying to herd her. She would walk a few paces, stop, dig her feet in, put her head down and no amount of tugging would get it up again. She seemed to think the whole idea was downright silly. I wondered how long it was going to take me to get her to Lickeen. I had a feeling that Elly and I might well end up taking the scenic route.

Kitty had no such reservations about being led, but then she had had years of training. On a good day, neither did Anlon. He had grown so tall that I had to stretch to get the head-collar on him. In the spring, he was going to a horse expert on the other side of Bantry for the first stages in his training as a riding horse. Watching him buck and kick in play with his mother in the field, I was extremely glad that this was one task I would not be called on to perform. My riding skills were much better suited to Kitty's slow and stately pace.

Ray was back at Lickeen and I was very glad to see him return. There still seemed so much to be done and if I was to make my moving-in deadline we didn't have a lot of time. He had nearly finished rebuilding the porch and I was helping him. When we stopped for a break, we would listen for the sound of the workmen who were currently stringing thick, power-bearing cables between the electricity poles. They'd had to spend several days hacking down dangerous tree-limbs before starting on this phase. Then they'd been called away to repair extensive storm damage to cables further down the Beara Peninsula. I couldn't believe it when they finished. The whole process had taken so long that I was fully expecting a freak tornado or a sudden snow-storm to prevent them from completing the work.

When one of the tired and dirty crew passed by my house on his way to their truck that afternoon, I asked, for the umpteenth time, 'How long?'

'Maybe tonight. It depends on the weather.' He cast a wary eye at the darkening clouds. 'If we don't get finished today, you'll definitely be on by tomorrow,' he added, trying to cheer me up.

They didn't finish that day and overnight the weather became worse but the following morning the work crew arrived early, clad in yellow oilskins. Just as it was getting dark, electricity finally came to Lickeen. A neighbour who had been walking up the Barley Lake road looking for wayward sheep had turned to face Lickeen at the exact moment the power was finally connected and the house and surrounding dark hillside were suddenly consumed in a blaze of light. He told me that he had stood stock still, amazed by this spectacle then said to no one in particular, 'Mother of God, will you look at that now? There's light where there was never light before.'

He couldn't have been more amazed than I was. I stopped what I was doing and rushed round to every light switch in the house, flicking them on and off delightedly. Elly deserved an extra-special treat for tea that night. After all, if it weren't for her none of this would have been possible.

Now when we came to work, no matter how dark and gloomy it was outside, we could get things done. We were able to make use of power tools for the first time too. This was a great help to Ray, who had finished the porch and had begun laying a new floor in the bedroom where my bath and inside toilet would soon be. I hadn't intended to have new floors. We had looked over the old ones closely, detected a few woodworm holes and uneven patches here and there, but it was no worse than you would expect to find in any old house. The floor-boards were sound enough and would do for now. Then, one day, I had been removing plaster upstairs, swinging my hatchet with a near-demented fury because I had become so sick of this process, when I hit a patch of wall near the apex of the old joists. For about a foot down from the ceiling the stones were loosely packed and had never been cemented in place. Ray said this was because the house had previously been thatched and the stones had been used to raise the height when the roof was slated some sixty years ago. In my pique I dislodged a huge boulder which narrowly missed my head before plummeting through the floorboards to the kitchen. It was obvious that this floor at least wouldn't do.

Today the kitchen was bare, swept clean of what, God willing, would be the last of the dust and cement. Its newly plastered walls still had that sharp, damp smell about them but they were drying out quickly and I would soon be able to paint. Drying the plaster had been all the excuse I needed to have both fires roaring all day. The small front room grate had a backboiler installed in it now and the radiators connected to it made the air in the house moist and warm. I would sit snugly on an old tree-stump in front of the kitchen fire while the rain beat against the windowpanes, thinking about how I would arrange the rooms and what colour I was going to paint the walls.

Then disaster struck. At the eleventh hour Ray came down with a bad dose of flu, along with half of the rest of the glen. I didn't catch it but my Christmas preparations were way behind, and I had to figure out how to get the house finished. Bryn and Kristine were not going to be able to make it for Christmas because of unexpected family commitments. But Amber would be here soon.

The bathroom still had to be built, and the cast-iron bath and my new inside toilet installed. Much of the ceiling upstairs still had to be insulated and plasterboarded. There were still several large holes in the walls made by the plumber when he was running his pipes. He had to return and connect the appliances in the bathroom – when there was one. Two panes of glass were missing from the back door and I hadn't done any painting yet. The last few days had degenerated into a blurred haze of running manically between the two houses, trying to measure unwieldy rolls of insulating material in one, chucking sundry objects into boxes in the other, and only managing to increase the level of chaos in both.

I made the tea extra strong when I got back to Barley Lake Cottage on the third day of this impossible schedule and sat sipping it amongst the wreck of my belongings, remembering the day-dreams I'd had back in the early summer – winter roses rambling round the door, a garden laid and magically blooming, and an overall feeling of well-ordered calm in my new home's freshly painted interior.

The next morning I had a lucky break. I was introduced to Sean, a local lad who worked in England and was home for the holidays. Sean was a builder, carpenter, general all-rounder and was amenable to the prospect of doing some holiday work. In addition to his many other talents he was gifted with animals and had a gentle, easy-going disposition. It was quickly tested to the full when he started working at Lickeen.

He patched, finished off floors, installed the insulation, and would happily go off to feed the animals. Suddenly there were only two days left before moving and only five days until the 25th. Sean worked around the clock while I slapped paint on the walls downstairs. Not once did his spirits flag or did he lose his temper. He just worked steadily on, determined that the place would be ready by the time the cattle truck I'd booked came to the cottage for my belongings.

On the last afternoon of this marathon stint, the people who were laying the floor-covering arrived. Originally, I had planned to have painting and other mucky work finished long before this event but I had totally given up on any thought of logical sequence. By the time they had left, the house was transformed, carpeted in soft heathery colours in the front room and upstairs, and on the kitchen and porch floors, a pale grey linoleum. The house looked beautiful – quiet, clean and waiting.

As the carpet-layers were leaving the plumber arrived. Sean had somehow managed to finish building the bathroom. Now the toilet, a small sink and the cast-iron bath-tub were ready to be connected to the pipes. The toilet and washbasin were no trouble. They were light, and Sean and I already had them waiting in the newly built room. The bath-tub was another matter.

It took four of us – the plumber, his mate, Sean and myself – over an hour to inch the bath into the house and manoeuvre it up the narrow stairs. This last stage in the operation involved tying ropes to it and hauling it from above. By the time we had finished we were hot, sweaty and exhausted. But the bath was finally in its allotted space and connected to my water supply. It was strange to think that one small pipe in the stream was going to supply the whole house, bath and all.

Sean and I worked on after the plumber left, stopping briefly to warm a pizza in the oven. We sat and munched great wedges of it, too tired for talk, conserving our energy for what was left to be done. At about ten o'clock that night, when we were back at work, the house was suddenly illuminated by car headlights sweeping up the boreen.

It was my farmer friend, Dermot, his small lorry piled high with hay and straw which I'd forgotten I'd ordered. He had been very busy himself and had been unable to get to me earlier. Sean and I took a welcome break and went to help him unload.

It was a clear, still night, no moon and a million or more stars twinkling in the sky. One day, I promised myself, I would buy a telescope and on a night like this I would take it out to the cliff which overlooks the house and watch the stars, learn something about the constellations. In the short walk from the house to the barn I had counted two shooting stars and I could see a satellite whizzing across the open night skies.

The bales of hay were full of the scents of summer. There were heads of clover, thistles and wild herbs in amongst the dried grasses, and they felt warm to the touch. Soon they were stacked neatly in a corner of the old barn, filling it with their aroma. It was amazing how the barn suddenly seemed to come to life. It looked purposeful, ready for anything now. I couldn't wait to bring the animals to Lickeen. I was sure they were going to love their new accommodation. But first, there was the small matter of my own house to worry about. Tomorrow I would be moving in.

I went back to painting. *I* was fading fast and starting to imagine things – I tried to grasp a door-handle that wasn't there, heard several voices outside and went to the front door thinking we had midnight callers. By about two o'clock in the morning, I realized that I had painted the same stretch of wall three times at least. In the rush, there had been no time to apply undercoat so I had opted for a good quality paint that boasted it didn't need one. Some patches did, however, and I had quickly lost track of those I had gone over more than once. I didn't care any more either.

I was beyond being tired and had entered the Twilight Zone some hours ago. As the sky began to lighten with the coming dawn I realized that, finished or not, we would have to stop soon. The man with the cattle truck was due at Barley Lake at 8.30 and I had hardly begun the packing.

CHAPTER SIXTEEN

HOME AT LAST

Never try to move just before Christmas, particularly if the new house was recently derelict. My own long-awaited move to Lickeen had all the makings of a fully fledged disaster. Amber was due home any minute and there were only mountains of boxes and heaps of scattered possessions to welcome her.

Wearily, I surveyed the chaos that was Barley Lake Cottage. My once reasonably tidy, compact home looked as if a gang of wreckers had charged through it. Sean, who had gamely come back to Barley Lake with me to help with the move, had promptly passed out on the couch, his mug of tea cold and forgotten beside him.

Theoretically, there is no reason why a person cannot move in an organized way – labelled boxes, neatly taped containers – that sort of thing. I once shared a house with someone who did just that and was open-mouthed with admiration at this woman's military-style precision when she moved in. I have moved many times but I have never been able to come close to this ideal.

This move was particularly gruesome. Blearily I searched

the cluttered kitchen table for a mug. I needed tea, and plenty of it. I found one that had been used for storing paper-clips, emptied them out and put the kettle on. The dogs lay outside the front door nose-to-tail looking depressed. They had decided that life as they knew it was coming to an end and they didn't want to think about what was going to happen next. I wished I could have told them that it was going to be all right, that we would soon be moved into Lickeen, which they loved. But just now I hardly believed it myself.

As I sipped my tea, I continued to put things in boxes and make mounds of possessions on any available floor space, straining my ears for the sound of the lorry coming up the hill. Sean slept on.

It was as well the man driving the truck was late. By the time he finally arrived I had one load ready. Transporting my possessions consisted of scooping up great armfuls of things and hurling them into the open truck until there wasn't any more room. After we had filled it, he trundled off towards Lickeen.

When I went back inside, the house didn't look any different. In fact, it looked as though there was more stuff here than before. At this rate, it was going to take all day. On the second or third run the driver was taking my small couch and the bedding that was piled on top of it. He yanked up an armful of blankets when there was a sudden, piercing yell. In his deepening sleep, Sean had burrowed into the blankets and been completely forgotten. He stretched, grinned at us, gulped down his cold tea and began helping.

Whatever difficulties the truck driver encountered during that long day, and there were many, he always had the same cheerful answer. 'No problem,' he would say, squeezing a wardrobe down the narrow staircase and through the equally narrow front door. 'No problem,' as he forced things round and over any obstacle in his path. By the end of that afternoon, each time we hit trouble invariably someone would say, 'No problem,' and we would all laugh hysterically.

It was just getting dark when we chugged up the boreen with the last load. My belongings were scattered everywhere. Furniture, boxes of books and clothes were dumped outside and would have to be brought inside before it got dark. The house seemed crammed to bursting point. I couldn't understand it. This place was bigger than Barley Lake Cottage. There should have been loads of room. But I had done it. I was finally living at Lickeen.

What I wanted more than anything was sleep, on top of or around any of the miscellaneous objects that were scattered about. Though I was unable to complete any sentence and had resorted to monosyllabic grunts, there was nothing to eat in the house and I was determined to feed Sean who was sitting in a corner by the kitchen fire.

It was dark and still when I stepped outside the front door, and very cold. The walls of my new home were three foot thick and the fires had been going all day. With all the rushing around I hadn't noticed the sudden drop in temperature. But my car windows had a thick, sparkling layer of frost on them and the old Renault was reluctant to start.

I drove groggily down to the village where I bought a bottle of wine, several Mars Bars and rented what proved to be one of the worst videos I have ever seen. Somewhere in the dim recesses of my mind was the realization that this night was special, an occasion that had to be marked by something more significant than another hunk of lukewarm pizza. I opted for egg and chips. We could eat and watch a video, the height of suburban sophistication. I completed my purchases and headed for home. Soon I would be able to sleep.

In front of the blazing front room fire, lying on whatever was to hand, we ate, drank large amounts of wine and drowsily tried to make some sense of the film. The last thing

I remember was a brassy, archetypal barmaid complete with plunging neckline trying to convince her husband that she had taken a lover for his sake and not hers.

The following morning I awoke feeling drugged and aching. Sean had, quite sensibly, gone home some time during the night, probably driven out by the terrible film. He had washed up and tidied away our supper things before leaving. I could have continued sleeping right where I was, curled up on a pile of rugs, for at least a week. But there was little chance of that. Amber was arriving tomorrow and this was my first day in residence. I wanted to make the most of it. Now . . . if I could only find the coffee and the kettle. I was in dire need of my morning pint.

But when I turned on the tap nothing happened. I wiggled it about a bit, turned it on and off. Still no water. I gave it a swift blow with a hammer, but it was no good. The water that had unfailingly supplied us throughout the summer refused to flow.

I opened the front door and stamped outside to poke around and see if I could find out what was wrong. The icy air hit me with a bone-numbing blast, but it was also one of the most beautiful mornings I have ever had the pleasure to witness. The frost that had formed during the night was several inches thick – crunchy underfoot, it looked like snow. The sky was Swiss Alpine blue and the sun shone over all this splendour with a blinding brilliance, sparking off thousands of tiny particles of ice with a dancing blue fire. On the very tops of the Caha mountains lay a fine dusting of snow and the fir-trees in Glengarriff Forest looked as if they had been cut out of a Christmas card and doused with glitter.

Overnight, everything had frozen rock hard, the car, the mud-bath outside the back door, some small puddles and, of course, my water supply. I would just have to wait for it to thaw. I pulled on a pair of wellies and reached for my voluminous overcoat. When I was suitably protected I found a bucket, called the dogs and headed for the stream. Sam and

Tom were, of course, delighted. Clearly they approved of the effect Lickeen had on me. This is the way to live, they seemed to say, plumy tails waving – wake up, don't waste time over dressing or coffee, and immediately go for a long brisk walk. We headed down the boreen to the stream.

When it eventually thawed that morning, further investigation revealed that the frost had been so severe it had blown apart the joints in the water-pipe. While I was waiting for the weather to warm up, I'd tried to make a quick run to Barley Lake Cottage for several small items I'd left behind. I'd also managed to forget my bed and I was looking forward to sleeping in it that night. If I rolled the mattress I could get it in my doughty little hatchback. I could always collect the base on another run. But the hill was a solid sheet of ice and there was no getting up it, so I drove home to Lickeen and began unpacking.

I had called the plumber earlier but, needless to say, he was busy after the severe freeze and there was no telling when he would be able to get to me. Sean was coming back later that day to finish off a few odds and ends so perhaps we would be able to get the water sorted out between us before Amber arrived. It was more than a little annoying to have an inside toilet and a bath at long last, but not be able to use them.

The next day, 23 December, I was going to Cork Airport to pick up my daughter. Apart from a few cards I had received, there wasn't a sign of Christmas about the place. I had bought a box of cards sometime at the beginning of December, and I'd had every intention of sending them out. I enjoyed Christmas and usually loved all the fuss and the preparations but I hadn't got round to writing so much as one card this year and it was certainly too late now. Neither had I done any baking.

Normally, at the beginning of the Christmas month, I gather together pounds of dried fruit, candied peel, brandy, put on a radio broadcast of Dylan Thomas reading *A Child's*

Christmas In Wales, and get baking. I make mincemeat, puddings and a Christmas cake with Guinness in it and enjoy sliding into the festive spirit. This year I had religiously bought all the ingredients but had got no further than that.

I went back inside and built up the kitchen fire until it blazed high up the chimney. The front room grate was stoked with a mixture of wood and sweet-smelling turf and the radiators were maintaining a steady heat which warmed the whole house. For the next few hours I sorted and unpacked. I hung my favourite picture on the kitchen wall and pulled out a few of my best-loved books to put up on the shelf. I unearthed several pairs of old cream-coloured lace curtains that I had bought at a flea-market some years ago and had never used. I had painted the walls a pale, lavender blue and the woodwork was a buttermilk shade. The curtains should match this colour-scheme perfectly.

I held them up against the window. It looked as though they would fit. It was still cold outside but the sun streamed through the small window-panes and the kitchen was warm. I didn't care about the unpacking or even the lack of water for the moment. It was enough to be sitting here before my own fireside.

I was at the stream with the bucket getting more water when Sean arrived. It was mid afternoon and he apologized for being late. He had been unable to wake up that morning. I could quite see why.

We decided to see if anything could be done about the water before it got too dark, and set off along the mountain path to the stream where the pipe was lodged. We cleaned out the filter on the end of the pipe and laid it back in the stream. Then we retraced the pipe's circuitous route towards the house until we came to the place where the joints had come apart, and put them back together. Now it was just a question of getting the water flowing again.

I remembered Ray telling me that he had used a pump, but as we didn't have one to hand, Sean thought perhaps we could get things moving by alternately blowing and sucking into the pipes. I was stationed at the end of the pipe nearest the house, Sean was up at the stream. I was instructed to suck at five minute intervals. This I did until my cheeks were red with cold and effort, but to no avail. Sean came back down the mountain shaking his head. 'I don't know if this is going to work,' he said, 'and it's getting too dark up there to see what you're doing. I think we'll have to try again tomorrow.'

One more night without a bath wasn't going to kill me. I headed for the warmth of the kitchen. I still hadn't got back to Barley Lake Cottage to pick up my bed but the hill was probably frozen again, and anyway I was much too tired and warm to think of stirring. That night I settled down to sleep on a spare mattress but in my own bedroom. There was a skylight over my bed and through it I could see hundreds of stars. The night was clear and cold and there was no wind. That probably meant another severe frost, but I was too cosy to care.

I had done it. I had made the deadline. What kind of a Christmas we were about to have remained to be seen.

CHAPTER SEVENTEEN

CHRISTMAS AT LICKEEN

Cork's compact airport is a bustling and happy place at Christmas time. There are huge signs welcoming the many expatriot Irish home for the holidays, fairy-lights festooning trees and in the car park, a huge crib complete with replicas of the Holy Family and attendant animals. Several large speakers are attached to the crib's thatched roof. As I pulled my battered Renault into a parking place and opened the door, the strains of 'Silent Night' filled the cold air.

The car had again not wanted to start that morning. I hadn't liked the sound of the engine, but eventually it had been persuaded into life. This delay had made me late and I dashed for the arrival terminal, skirting families pushing carts piled high with gifts, and relatives anxiously scanning the flight information board to see if there were any delays. Then I spotted Amber, surrounded by luggage and packages, smiling and waving to me.

We decided to have a much-needed cup of coffee before

starting out on the two-hour journey home. Standing in line at the counter of the cafeteria upstairs, planning how I was going to break the news about the lack of water, I happened to glance down at my hands. They were absolutely filthy. Not the kind of dirt you might acquire after a couple of days' graft, this was of the deeply ingrained variety, as if I had spent most of my life working as a car mechanic. How had I not noticed? It wasn't that I hadn't washed, scrubbed even. Suddenly I felt shabby and self-conscious amongst all these people in their holiday finery. I curled my hands tightly around the tray so that less of them could be seen. A couple of nights ago I had tried to scrape some of the residue of tar, soot and smoke off the kitchen beams with a wire brush and sugar-soap. The stiff brush had made holes in the gloves I was wearing and the gooey black mixture that came off soaked straight into my hands which were especially porous after weeks of working with cement. I paid for our food gingerly and scurried back to the table.

As we drank coffee and ate huge sticky Danish pastries, I filled Amber in on what had been happening, how I had managed to get moved in but only just. I told her about the water. There was little point in trying to hide anything. She simply smiled, said how glad she was to be home, and congratulated me for having succeeded in moving in at all.

I began to feel slightly queasy during the drive home. I was probably just worn out from the last few days. I still hadn't managed to catch up on my sleep and I knew I had drunk too much coffee that morning. I could slow down and relax a bit now the holidays were here.

That afternoon Amber, Sean and I went up to the stream to try to get the water going. Once again, the task of sucking on the pipe fell to me. Amber and Sean would be at the end which was in the stream, taking turns to blow.

My cheeks hollowed with effort and I was feeling sicker by the minute. Nothing happened. I tried again. This time a rush of brackish water shot down the pipe and filled my mouth. It left a horrible metallic taste after I spat it out. The flow stopped as abruptly as it began.

We tried again. It was still no good. We would get a sudden gush of water then nothing. By this time, I was beginning to feel decidedly ill. Fortunately, my co-conspirators came bounding down the mountain just then, for I don't think I could have kept up this performance much longer.

Amber looked at me suspiciously.

'What on earth's wrong with you? You look terrible.'

I smiled weakly. 'It's probably all caught up with me. Actually, I feel terrible. I think I'll go and lie down for a while.'

With that I lurched off towards the house, leaving them to manage as best they could. For the next twenty-four hours I was completely out of commission. I was wretchedly ill, sick from a combination of overwork, not having eaten properly and insufficient sleep. I retched for what seemed like an eternity and in between bouts of that I slept. It wasn't much of a welcome home for Amber.

She and Sean fed Kitty, Elly and Anlon and saw to the dogs. They even went and picked up my bed from Barley Lake Cottage in Sean's car. Ray, who had recovered from the flu, had stopped by to visit, and with his help, they got the water running again. I knew nothing of all this as I lay in the darkened bedroom wishing I were dead, nor did I care. I didn't even have the strength to say, 'Bah, humbug.'

By two o'clock on Christmas Eve I was up, sitting in a chair by the roaring fire feeling weak and light-headed but over the worst of it. Tomorrow we would be preparing Christmas dinner in this kitchen. I looked round me blankly. The trouble was, we had virtually no food in. Since Amber was not insured to drive my car and I could hardly ask Sean, after all he had done, to go shopping for me on Christmas Eve, there was only one thing for it. I would have to get to

Bantry and try to condense what would usually be several weeks of preparation into a few fun-packed hours.

I don't remember much about that shopping trip, just a confused blur of happy faces, Christmas carols, and shelves that in another hour would be completely empty. Bantry was packed with people shopping with the kind of fervour you'd expect from a crowd who had just learned there would be no more food – ever. I gave up working out what we needed and just bought everything that was left. It seemed simpler. And I wanted to get home.

We gave up the unequal struggle to create order out of chaos inside the house and before total darkness descended, Amber and I dashed outside with the bow-saw. We fell upon a handy fir-tree and sawed energetically for a few minutes. The pungent scent of pine-sap rose sharply on the damp air. When we had cut through the trunk we stood silently for a moment looking over the deep, dark green of the forest to the Caha mountains beyond. The first stars were just becoming visible above them. It was a moment of perfect serenity.

We dragged our prize indoors. Suddenly it began to feel like Christmas. It was a fine tree, tall and upstanding and its festive spirit filled the house. I threw the largest log I could find onto the fire and without further discussion we settled back and started on the duty-free.

Several days later we were still doing pretty much the same thing although I would vary the pace every now and then by having long, hot baths. This was the biggest treat of all and we had made an important discovery. The huge open fire in our kitchen was no ordinary fire. To feed it even once was to become its slave, caught up in the magic of its flame-pictures. Hours passed where we talked about nothing more significant than the merits of different types of wood and the best combinations of species for creating the most satisfying blaze. These intense discussions were interrupted only by one or other of us wandering off in search of the ultimate log for that evening, or to ferret out the odd shop-bought mince pie. Videos we had rented went unwatched and even the radio was, for once, ignored. The fire, and its care, was entertainment enough. It was a unique and special Christmas and one which neither of us is likely to forget.

Before Amber returned to England and college, I decided to enlist her help in bringing Elly to Lickeen. I had intended to do this before Christmas but, what with one thing and another, I just hadn't got round to it. I was still worried about having Anlon on this rough land, so he was going to be spending a few weeks with a neighbour's pony in a nice flat field. The colt he would be staying with was the same age, and shared Anlon's lunatic energy level. Grazing large animals, particularly in the winter, requires constant shifting and juggling if the field is not to become poached and the grass made useless for the following year. I always seemed to be moving mine around.

Elly was glad to see us. She was back in a neighbour's field near Barley Lake and thought that she was about to get an unexpected mid-afternoon snack. When I opened the gate she hurtled out and promptly buried her face in a nearby clump of grass refusing to move. I had to tap her several times with the stick I was carrying before we could get her going. Apart from a few unplanned detours into fields whose gates were open, we managed pretty well. Elly seemed to know that something was afoot. When we reached the bottom of the Barley Lake road where you turn towards Lickeen, she stopped and sniffed the air suspiciously. It was further than she had ever been before. Amber went ahead of us and opened the gate. Elly took the boreen at a fast clip as if she couldn't wait to get home. Once at the house she had a long drink of water from the bucket I'd filled for her, then flopped down happily on a deep bed of straw in the barn.

After years of working alongside humans in the forest, Kitty was more used to the company of people than that of animals. The next day I watched her as she sniffed curiously, ears pricked forward and an alert expression in her deep dark eyes. She knew there was another animal at Lickeen. I took off her head-collar and Amber gave her a carrot. Then Kitty set off to explore. Elly was in the field behind the barn. Amber and I watched as she stared in Kitty's direction, a faint look of contempt on her haughty features. Kitty held her gaze for a few moments then busied herself by munching on a nearby holly bush. I wondered how Elly was going to react to Kitty's presence in *her* barn tonight.

Amber's holiday came to an end that day. We both agreed that it had certainly been different as we hugged goodbye sadly at Cork Airport. Threatening dark clouds blew up as I was driving back from the airport and there was a cold, cutting edge to the wind. It was definitely a night for being snugly tucked up.

When I called Elly for her feed later, she trotted towards the barn as fast as her growing bulk would allow. She seemed glad to get inside and out of the cold and knew there would be hay and a nice bed of straw waiting for her. I pulled my jacket tighter around me, anxious to finish my chores and get back to the fire.

Kitty ambled round to the back of the house and stood looking enquiringly into the barn. I decided to lead her in, feed her, then leave her to come and go as she pleased. Elly was safely bedded down in her own pen in the corner so they would be able to get used to each other from a safe distance.

Kitty stood in the barn doorway blocking the last remaining light. I waited beside her letting her set her own pace. The look of amazement on Elly's face as she stared at Kitty was downright comical. She swung her head round so that she could see this intruder better, snorted, then slowly heaved herself up onto her feet and glared. When Kitty failed to respond to this challenge, Elly let out an enraged bellow the likes of which I had not heard since that first terrible night she had come to me. With the grace of a ballet-dancer Kitty stepped delicately into the barn, totally ignoring the rude and angry heifer. She went over to the hay-net and began sifting through its sweetly scented contents looking for the choicest morsels.

Elly, furious that her full war-cry had had no effect, lowered her head and charged at the fence. The sight of this self-possessed horse calmly eating was too much for her. Kitty ignored her. Elly looked perplexed as if wondering what in her bag of tricks she should try next. But the lateness of the hour and her own bulk overcame her territorial tendencies and she settled down to sleep, pointedly turning her back to Kitty. I considered this an opportune moment to leave them to it.

By about one o'clock that night a fierce storm had blown up. The wind sobbed and moaned so loudly around the house that it would have been impossible even to hear Elly bellowing. I couldn't have been asleep for more than an hour when a loud crash woke me. I jumped out of bed and hurried downstairs fearing the worst, convinced that the sudden noise had something to do with the two recently introduced barn-mates.

The back door stood wide open and a gale was blowing through the kitchen, sending the smoke from the fire spiralling crazily out into the room, and blowing the papers on the table in all directions. For one far-fetched moment I thought that Elly, furious with me for landing her with a room-mate, had broken out of her pen and charged the back door, determined to teach me a lesson by taking up residence in *my* house. I wouldn't have put it past her.

I peered apprehensively around the shadowy kitchen, half expecting her to jump out at me, but nothing stirred. Relieved, I concluded that the wind must have blown the door open. I searched groggily for the key and found it under a pile of well-matured bills. I would have to lock the door against the wind. The barn, I noticed with relief, remained silent.

Apart from the scattered debris of leaves and branches,

the next morning dawned fine. The sky was a clear, washed blue after the storm and there was no sign of rain. Kitty was standing with her head poking out of the barn doorway. She greeted me with a friendly whicker. Elly mooed reproachfully when I went inside but, apart from that token protest, she seemed to have become resigned to co-habiting. I tried to settle back into some sort of routine in the days following Amber's departure but it wasn't easy. I was still on a high from Christmas, the move and the heady sensation of finally living at Lickeen. Eventually I managed to strike a fairly even balance between writing, sorting out the house, and the on-going task of caring for the animals in winter, when they must be constantly watered and fed.

Characteristically, the year had begun with a mixture of good and bad news. The stories I had written about Lickeen for the paper and, more recently, RTE Radio, were getting an increasingly good response. The bad news was that the bills were piling up again. Several friends had suggested that I wrote a book about the Lickeen experience and my whole *Carry On Farming* approach. I thought it a good idea, but I couldn't see how I was ever going to find the time.

CHAPTER EIGHTEEN

HARD LABOUR

March, when Elly's calf was due, was nearly upon us and I was getting distinctly nervous. Not for the first time I wished I had a more accurate date.

The first calving of a heifer is tricky at the best of times and is usually closely monitored even when you know the due date and the bull involved. At this time of year I would often meet neighbours haggard from lack of sleep after checking a near-calving heifer every three or four hours until she finally obliged. In another few weeks it would be time for me to begin my own vigil. I had read all the books, talked to my neighbours and gleaned as much information as I could about the calving process, paying particular and morbid attention to any problems that might occur. I had decided that this birth was going to be no time for heroics.

All over the glen right now calves were being born unremarked, but their owners had been through all this many times before. The vet was only called out in an emergency. Normally, you managed by yourself. One of the most common problems was a calf getting stuck at the shoulders. Its owner would then have to help ease its passage into the world. If muscle-power failed then it was ropes and jacks. I winced every time I thought of them.

Elly continued to thrive despite my anxieties. She was round and fat and her coat was a glossy red. I was coming to enjoy standing around looking at my animals, assessing their condition, seeing how they communicated with each other, how they attempted to make their wishes known to humans who must sometimes have appeared pretty thick to them. I particularly approved of the way cows took the business of relaxation with a determined seriousness, settling down for prolonged periods of cudding in the most comfortable spot they could find, legs tucked neatly beneath them. They would chew and doze happily for hours, unlike horses who usually sleep standing up. Often when I was watching Kitty have a nap, I wanted to go and shake her, wake her up and ask her why on earth she didn't lie down comfortably instead of shifting her massive weight from foot to foot as her head sank lower and lower? She looked for all the world like some old drunk about to pass out in a doorway.

Elly no longer raced across the field for her feed. Now she came towards me at a slow and matronly pace. Once the drama of this birth was over I was going to have to learn how to milk. I still hadn't figured out how I was going to handle this daily event. First, I had never done it before and second, perhaps more significant, neither had Elly. We would not be a winning combination in this venture. Some people delighted in telling me – usually with ill-concealed sniggers – that heifers could be hard to milk, particularly those who had come to think of themselves as large dogs. The idea was that Elly's calf would have all the milk it wanted from her during the day, then at night I would separate them and milk Elly in the mornings. In theory this plan seemed both practical and fair. But I already knew enough about animals to realize that it was not going to be that straightforward. I would need time—and time was more important than usual just at the moment: I had finally taken heed of the advice and encouragement and started writing a book about the Lickeen Experience.

The economic forecast continued gloomy. No matter how many feature stories I wrote, I still couldn't bring in enough money. And I was becoming increasingly concerned about my old Renault. When I tried to start it, the noises coming from the engine were not pleasant and I did my best to ignore them, although I had a sneaking suspicion that it was something to do with the starter motor.

I met a cattle-dealer friend in the village one day when I was trying to get the car started. He knew about my money worries and he offered to sell Elly for me. She would fetch a good price at the market right now, he assured me. Elly was in great condition and she was heavy in calf. And I could always get another cow later in the spring. It was a reasonable suggestion but I could hardly answer him. I mumbled my thanks and said something about thinking it over. Surely it wouldn't come to that? Elly, used to lots of attention and the occasional treat, going to a normal farm where she would probably have a number stamped on her backside and be prodded with sharp sticks? She would be devastated. So would I.

In the middle of these gloomy speculations I'd received a letter from Amber. She was suffering from an attack of winter blues. She had flu, the city streets were grey and dirty, the weather bitterly cold, and she wished above all things that she could be at Lickeen for the first signs of spring. The optimistic green shoots of daffodils which carpeted the sloping ground at the front of the house were already pushing sturdily up through the dark soil, and several trees had tightly-furled bright green buds. Spring-time at Lickeen was going to be spectacular.

Elly had begun behaving strangely, staring vacant-eyed into unseeable distances. One day I found her standing bemusedly in a very overgrown part of the stream, bellowing distractedly and unable to get herself out again. I had had to run back to the house for a bow-saw and slasher and cut a path through a thick clump of rhododendrons to release her.

My sleep was punctuated by the unfriendly ringing of the alarm-clock, set to go off every four hours. I would fall out of bed, pull on as many layers of clothes as I could fit over my night-wear and, armed with the storm-lantern, make my way out to the barn. After the fourth night of this nerve-wracking vigil I was fading fast and we seemed to be no nearer the main event. It was Saturday night and the couple who had bought Anlon were at Lickeen for a visit. They were articulate and witty people and I always enjoyed their company. But I wasn't exactly the life and soul of the party that night, yawning profusely despite my best efforts and feeling as though I was in distinct danger of falling asleep face down on my plate at any moment. I would excuse myself at regular intervals and go outside into the freezing night air to check on Elly. Surely it wouldn't be much longer now?

When my guests left I went to bed for a couple of hours, until it was time for the next trip to the barn, certain that there would be nothing doing tonight – perhaps ever. Elly too, was obviously undergoing some sort of phantom pregnancy.

It must have been about four the next morning when her first enraged bellows shattered my sleep. This time, I didn't bother with the usual layers of clothing. I just grabbed my overcoat, shoved my sockless feet into my wellies and ran for the barn. This was it.

Elly was well into labour. She strained and groaned and bellowed angrily. I took a few deep breaths to calm myself as I hung the flickering storm lantern on a nail in the beam. I had a strong urge to run back to the house, start boiling water and tearing up sheets but somehow I managed to control myself. There was nothing for me to do at the moment but watch and wait.

Elly would lie down for ten minutes, her flanks would heave, she would bellow a few times, then she would get up and pace around her pen again. She kept looking at me and I had the feeling she was glad I was there. Finally I saw the white, membranous sac containing her calf appear for a couple of seconds before vanishing again. The suspense was awful. But after a couple more gigantic heaves, Elly gave birth. I watched the first tentative limb emerge through the sac, and suddenly there he was, a fine-looking bull calf, all in a heap at Elly's feet. I couldn't have been more delighted. She had managed just fine on her own.

I decided to bring my morning shot of caffeine into the barn and enjoy the new arrival's first few hours on earth. Five minutes later, armed with another sweater, some socks and my mug, I settled myself comfortably on the hay. The calf was up on his feet and tottering around now. He was the same deep red as Elly and he even had an identical white blaze on his forehead. To all outward appearances he seemed fine. But I couldn't seem to rid myself of the nagging feeling that all was not well. As yet, he had made no attempt to suckle. In fact, he had shown no interest at all in Elly's bulging udder.

I watched and waited. Despite Elly's increasingly frantic urgings and persistent lickings, the little calf just stood in one corner of the barn with a bemused expression on his small face. Then he suddenly sank to the ground as though his legs could no longer bear the weight of his small body.

I put my coffee cup down, went cautiously into the stall and piled straw up around him until he was almost covered. I reached out a hand and stroked him gently. He felt cold and made no attempt to suck my fingers when I held them out to him. I went inside to consult my books.

Some calves can take time to come around after birth, one book said, and although most will begin to suckle immediately, others can take up to an hour. I glanced at the clock. It had been forty-five minutes since he was born. In such cases, the book advised keeping the calf warm and attempting to get it suckling by squirting some milk onto your fingers and placing them in its mouth. The vital first milk often did the trick of reviving a calf that was weak after birth.

I hurried back to the barn, lugging the portable gas-heater and some newspapers to stuff up cracks in the wall behind Elly's pen. Soon the barn was so warm that I had to take off my overcoat. Elly edged away from me nervously when I approached her with a jug in one hand. Her calf was still lying in his nest where I had left him and he looked up at me with large, dark eyes.

Eventually I managed to persuade Elly to stand still. Her udder was taut and full and, inexperienced as I was, it was difficult to get anything out of it. But eventually I managed a thin lining of dark yellow colostrum on the bottom of the jug. I stuck my fingers into the stuff and put them under the calf's nose. He blinked and looked puzzled. I pushed my fingers into his mouth, feeling the first stirrings of real panic. He made no attempt to suck my fingers. I rubbed him frantically with some straw but despite this and the heat in the barn he remained cold.

Perhaps he was simply taking a long time to recover from the trauma of birth as the book said, and in another few minutes he would start to revive. I felt suddenly appallingly aware of my own ignorance. It was time to call for help. Several neighbours had offered their services. One of them might know what to do.

I quickly piled more straw up around the calf and made sure that the gas-heater was safe. Elly watched me as I moved around the barn. She looked anxious and miserable too.

When Sam has one of her large litters, there is always one that feels cold and has no suck reflex. I try everything I can to revive it, hot water bottles, holding it onto Sam's nipple.

But no matter what I do, that pup always dies.

CHAPTER NINETEEN

THAT'S FARMING

It was only half past seven when I called the first neighbour. He had just come in from checking a cow of his own who was due to calve at any moment. I told him about the events of the night and how concerned I was about Elly's calf. He said he would be right over.

I called the vet. He sounded sleepy but I explained the situation and he said he would be there as soon as he could. He lived on the other side of Bantry and it would probably take him the best part of an hour.

I hung up the phone and went back to the barn, afraid of what I might find. The calf was lying where I had left him with that same, bemused expression on his face. I picked him up gently and put him to Elly's bulging udder, placing a full, fat teat in his mouth and squirting a little of the lifegiving milk down his throat. It simply trickled out again.

My neighbour's tractor shuddered to a stop as he pulled up in front of the house.

'He doesn't look too good, does he?' he said, standing in the barn doorway and shaking his head. Just before he had come in, the calf had started bellowing in a surprisingly full-throated, adult tone. I had seized on this as a ray of hope that he was getting stronger.

'How long's he been making that noise?' my neighbour asked. He told me that it was a bad sign and that normally a calf would be much quieter. My spirits sank even lower.

He climbed into Elly's pen and examined the little animal all over. He could find nothing obviously wrong. Then he milked Elly into the jug and tried getting some liquid into the calf. Most of it trickled out again and the calf still made no attempt to swallow.

'It could be pneumonia,' my neighbour said worriedly. 'Newborns will sometimes get that, but you're keeping him warm. There's nothing else you can do until the vet gets here.'

We stood and watched the calf for a few minutes but there was no change. He continued to bellow piteously. My neighbour had to go back to check on his own cow and as I walked him to his tractor, the sound of the calf's cries followed me.

I kept rubbing him with straw while I waited for the vet. Although I had piled him up with blankets, he felt no warmer.

When Finbar arrived, he took one practised look at the ailing infant and said he didn't think the calf was going to make it. Despite this gloomy prognosis, he set to work anyway, and for the next two hours he tried every combination of injection, antibiotic and stimulant he could think of in an attempt to save the calf. Eventually, he admitted that he was as puzzled as I was. 'It doesn't make sense,' he said, scratching his head. 'He's full-term and full weight and there was no trouble with the birth. I don't think he's got pneumonia either. There's nothing obviously wrong with him. Perhaps it's internal, organs not properly formed, something like that? All we can do is to keep trying.'

When everything that could be tried had failed, Finbar decided to try a blood transfusion directly from Elly to her calf. Looking back on it now, this was the worst moment of the whole horrible experience.

Finbar handed me a huge pair of pincers which he said had to be clamped to the fleshy part of Elly's nose. I was to hold her while he inserted the tube that would carry the blood between them. Elly was completely distraught and her udder was bulging painfully. As I did what I had to do and clamped the callipers on to her nose, I muttered a useless apology under my breath. Elly did everything she could to shake me off but, somehow, I managed to keep hold of her. Blood from her nose was dripping into the straw.

Finbar finally managed to find a vein in the calf that hadn't collapsed, and the blood from his mother began to gush into his little body. He responded not at all.

'I'm sorry,' Finbar said, straightening up from his efforts and starting to put his instruments back into his bag, 'but I don't think he's going to make it. If he gets through the next hour, there might just be a chance, but don't get your hopes up. The only way we'll know what was wrong with him would be by doing an autopsy. I don't know that you'd really want the expense of that, would you? It's a first calving, she's a young heifer, and I'm afraid these things do happen. All the same, it's bad luck for you, losing your first calf.'

I thanked him for all his efforts, and said I'd let him know about the autopsy. Right now I was too distracted to think. At least I knew we had tried everything. Finbar had been working on the poor little scrap for over two hours. I asked him how much I owed him.

'I shouldn't really charge you anything. I think he'll die,' Finbar replied.

But that was hardly his fault. In the end, he agreed to take ten pounds, a ludicrously small amount for all the time and effort he'd put in.

After he'd left, I went slowly back into the barn knowing that it was just a matter of time. Elly had settled down next to her calf in patient vigil. She no longer seemed agitated, merely resigned.

Oblivious to the condition of the pen which was full of the evidence of birth – trampled, bloody straw and mire, I sat down beside them and placed the calf's head on my lap. His eyes had begun to glaze over. I stroked him gently and, about five minutes later, he died quietly with that same bewildered expression on his small face.

We sat there like that, Elly, the calf and me, for I don't know how long. It was strangely peaceful in the warm, dark barn. At one point, I remember glancing over at Elly and seeing several large tears, the kind that cows routinely produce, rolling out of her eyes and plopping gently onto one of the blankets that covered her dead calf. It was a coincidence I could have done without.

I decided not to have an autopsy. Apart from being upset, Elly was fine and there was no reason to suppose that anything like this would happen next time. I was going to have to bury the calf though. At the moment, it seemed as if Elly intended to keep vigil over his body for ever and I wasn't sure what she would do if I tried to remove him. There was also the question of her painfully bulging udder. I had assumed the calf would take care of that, at least until I had learned to milk. I was going to need some help.

I remembered someone locally who had recently lost a much-loved dog that he had buried with a simple and touching dignity. I would go to see him.

He listened sympathetically to my unsteady request for help, then reached for his old coat and a pick-axe. Between us, we distracted Elly, lifted the small body into a wheelbarrow and dug a grave, high up on the mountain. The dull thud of the pick-axe as it ploughed through stones and into the thin, rocky soil echoed mournfully around the glen. I was glad I wasn't doing this alone. We placed the calf, still wrapped in one of the blankets, carefully into the ground and covered over the small grave with stones.

My phone rang constantly that day as neighbours who'd heard the news called to say they were sorry. It was as if I'd had a death in the family and I suppose in a way I had. Although this was the sort of fatality that often occurred amongst livestock, they were obviously upset for me, realizing that not only was it a financial loss, but that I had made the fatal mistake of turning my animals into large household pets.

There wasn't much time for grieving though. I had to sort out the new and pressing problem of what to do about Elly's milk. I called Dermot, who had bought Elly for me. He said the best thing I could do would be to buy another calf right away and see if Elly could be induced to accept it. Meanwhile I was to get as much of the colostrum from her as I could. If I didn't, he told me her bag would become hard and sore and there would be the danger of mastitis. He would be at the cattle market later and he would see if he could find a suitable calf for me.

When I went out to Elly, she seemed agitated and moved around the barn restlessly. I felt tired and emotional myself. I did my best to milk her but it wasn't easy. I could only pray that tomorrow a calf would arrive to save us both.

Dermot called back the next day. The prices on calves were very high at the moment, so he had gone to a neighbour with a three-week-old heifer for sale. Knowing the urgency of the situation, he had gone ahead and bought her. They would be over within the hour. I breathed a heart-felt sigh of relief.

She was a compact and composed brown and white calf who trotted calmly out of the box when Dermot let down the ramp. We shepherded her into the barn and Elly, predictably, began bellowing furiously. When we put the calf in her pen, Elly promptly tried to flatten her against the wall. The calf was obviously hungry and becoming increasingly desperate to suckle. But every time she approached the tantalizingly full udder Elly charged and tried to mow her down.

When I'd called Dermot he'd told me to save the after-birth. If it were placed over an alien calf the cow might be persuaded to accept it. He had instructed me to store this large, smelly object in the manure-heap where it would keep fresh, which I had reluctantly done. Now I went to retrieve it.

As it turned out, this was an unpleasant and entirely pointless operation. Elly still wouldn't take the calf and continued to try to murder it with full body-slams against the barn wall.

Dermot's wife Helen had been watching this performance from the doorway. 'Do you have any perfume?' she asked me suddenly. I was a little taken aback by the question, especially at a time like this, but I said I had, and wondered what was coming next.

'We had a good few orphan lambs last year,' she explained, seeing the puzzled look on my face. 'We'd a lot of success getting ewes to accept them by using perfume.'

By now I was completely fascinated.

'What we did was to spray perfume on both ends of the sheep then put them together. Because they smelt the same the ewes accepted the lambs, no bother.'

I am not big on perfume but occasionally I indulge in a bottle of Coco Chanel which I love. I'm inclined not to save it for special occasions but to splash it on at random when I feel in need of a morale booster. The last time I had bought any was over a year ago. Having been in need of many uplifts since, there was only a small amount left. I went upstairs and found the bottle, happy to make this great sacrifice if it would help solve the impasse in the barn.

Both cattle were dutifully sprayed and the barn soon smelled like a whorehouse. Clouds of exotic Chanel emanated from the animals' warm hides. I could have sworn Elly was preening herself. We came to the end of the bottle, but Elly still hadn't calmed down.

'Well, you'll just have to get another bottle in Bantry,'

Helen said seriously. 'You can't change scents on them now you know.'

I stared at her in amazement. Did she have any idea just how much this stuff cost? I imagined adding 'Chanel – for cows' to my farm records. I could just picture the taxman's face.

We went inside, leaving the cattle to see if they would sort things out themselves, which is often the way with animals. While we were having a well-earned cup of tea, Dermot and Helen told me the story of how my new calf had been bought.

'It was because of your articles that I got her for such a good price,' Dermot said, reaching for the teapot.

He'd told the man from whom he had bought her that a friend of his had just lost a calf and needed to buy one quickly. He mentioned my name although these were not people I knew. But they had been following the Lickeen stories in the paper. 'You don't mean to tell me Elly lost her calf?' the woman of the house had said, apparently deeply shocked. She'd insisted that her husband sell me the calf for a good price because she knew I would be upset and that I'd had a lot of expense lately. There was, however, a condition attached to the sale, Dermot said, trying to stifle a grin. The previous owners wanted me to name the calf after one of them – either Patrick or Joan. I couldn't see myself out in the fields calling, 'Joan.' In the end, I settled for Paddy even though she was a heifer. It seemed to suit the robust little creature.

Elly could not be induced to let Paddy suckle from her. All that day I kept creeping back into the barn to take a look. It was still the same story. Although popular opinion supported the theory that eventually they would sort things out, I was afraid that Elly was really going to hurt the calf if she continued knocking her about. Poor Paddy already had a couple of grazes on her head.

There was only one thing for it. I would have to bucket-feed Paddy, who was getting hungrier by the minute. Ready or not, I was going to have to start milking – and right away.

Chapter Twenty

THE RELUCTANT MILKMAID

Neighbours gave me practical demonstrations, advice, handy hints. I had read all the books. Despite this outpouring of help, the first time I went warily into the barn carrying a small bucket, I felt hamfisted and awkward. I took the precaution of tying Elly up. But no matter how I squeezed and tugged, I couldn't seem to get a flow of milk going and my arms and shoulders ached unbearably. It took over two hours and several breaks (during which time I went outside, swore loudly and kicked a few rocks) to half fill that bucket. I kept looking at it, sure it must have a leak somewhere. This pathetic offering wasn't even enough to keep the calf happy, let alone fulfil my dreams of an abundant supply for the house.

For days I went round practising the recommended hand movements, making strange passes as if I were about to perform some magical illusion. You had to grip and squeeze so the book said, from the little finger up and with both hands consecutively. God knows, I tried, but it was over a week before I could co-ordinate anything more than two, tentative fingers.

It wasn't much fun for Elly either. She was feeling sorely put upon by being expected to be an instant mother to this strange and pushy little calf. Having me hacking and sawing inexpertly on her teats twice a day was doing little to improve her mood. I woke up in the mornings feeling as though my arms had been dipped in starch. Writing was agony and, on several occasions, Elly and I had heated skirmishes during milking which involved much raising of voices and racing round the barn. But eventually, slowly and surely, as everybody had said I would, I got the knack. And then, miraculously, milking started to become a pleasure. The warm yellow stream gushed into the bucket. I loved Elly's grassy, mountainy smell in the close confines of the small barn, the utter peacefulness when she settled down to be milked and there was no other sound but her steady breathing and that rhythmic flow. Soon I had lots of milk twice a day, and had come to look forward to these quiet sessions with Elly as a welcome relief from the hectic pace outside.

I still felt sad when I passed by the little pile of stones which marked the grave of Elly's first-born. But soon the breeding season would be upon us. Elly would come on heat and it would be time to try again. In the end, that's all you can do – try again. A farmer friend who called after the calf died had said to me by way of condolence, 'Denise girl, that's farming.'

The first exciting signs of spring were becoming visible everywhere I looked – bright, impossibly green new growth on trees and bushes, and, up on the mountain, dark, mysterious little pools that were suddenly alive with frog-spawn. From all around the glen now I could hear the continuous cacophony of bleating sheep about to give birth, in the process of giving birth or, after their lambs had been born, crying anxiously to them day and night warning them of prowling foxes, straying dogs, or other pitfalls that can befall an innocent new lamb. Although I am not that fond of sheep, every year I enjoyed the spectacle of those giddy and rapidly strengthening balls of fluff careening madly about the fields, playing wild games of tag, or suddenly jumping straight up in the air as if overwhelmed by life and all its golden potential.

I sat at my desk one blustery spring morning going through my diary and checking dates. There was a sense of urgency, regeneration all around me now. Sam, to the undisguised delight of her partner-in-life Tom, had already been on heat and was due to whelp around Easter. And it would soon be time to get Kitty back to the stallion, to try again for a foal.

This time, I wanted to see how she would fare with a thoroughbred. Kitty's sturdy breeding had meant that with the Irish draught stallion which had sired Anlon, she had produced a colt that, although heavy, was nowhere near his mother's statuesque proportions. With a thoroughbred, the product would be a lighter, faster horse, and if it were a filly . . .

This was where I had to stop myself. I could never bear the thought that Kitty would one day no longer be a part of my life. If she were to have a daughter, it would make this inevitability much easier to face. I made a note next to the date when I expected Kitty to come on heat.

My earlier qualms, when we were still doing up the house, about letting Kitty wander in and out at will had proved well-founded. Leaving a door open at Lickeen was a big mistake. Kitty would seize any opportunity to get inside the house. She would sidle into the porch then just stand there not doing much, filling the small space. When she calculated that you had probably forgotten about her, with a stealthiness that completely belied her size, she would slip into the main kitchen and head for the fruit bowl.

I had found her asleep on the porch one day when I wanted to paint it. I'd had the sink put there to make more room in the kitchen. Now Kitty was using the sink for a headrest as she dozed. She looked so comfortable I hadn't the heart to disturb her. I worked around her, painting through her legs and reaching up over her neck. She didn't stir once.

Paddy had to be kept in the barn for several weeks yet, so Kitty and Elly had the land to themselves. They got on well enough. Usually they tended to ignore each other. But one day I happened to be glancing out of the front room window when I saw Elly slip round to the front of the house where Kitty was snoozing. Elly took a good long run at her and head-butted her firmly in the rear. Kitty jumped – who wouldn't when so rudely awakened? – flattened her ears, then tore off down the boreen at a mad gallop. Elly smugly assumed Kitty's spot in front of the house, an expression of complete innocence on her face.

I should have been at the typewriter but spring was in my veins too. All morning I had pottered around the house, watched the animals for a while, anything but buckle down to the writing I should have been doing. The truth was, I wanted to be outside too, walking, looking at the teeming, busy world around me returning to life.

A half-hour's ramble up the mountain wouldn't do any harm, and I would come back refreshed, revitalized and ready to settle down. I had several stories due for the newspapers I regularly wrote for, and something for RTE Radio which I was to record in their Cork studios the following month. There was also The Book. The house and its many needs were having to take second place while I tried to get my career back on to some sort of sensible, income-producing footing.

An author friend of mine who has a house nearby came to visit with her editor. She was most encouraging, asked me to write a couple more chapters and send them to her. I had managed to finish five, enough, I hoped, to convince her, and I'd sent them out with many a prayer to guide them on their way. What really kept me sane and believing in The Book was the favourable response I got whenever the Lickeen stories appeared, incredibly kind letters from total strangers, congratulating me on what I was trying to accomplish and enquiring about the animals.

After I'd had my half-hour's skiving off, I would do something useful and go down to check the mail. Perhaps there would be some response from the publishers. The dogs began to dash around in ever-decreasing circles, realizing, as I pulled on my socks and wellies, that an unscheduled walk was in the offing. Sam's natural enthusiasm was, however, a little inhibited because she had ballooned considerably in the last week. Dogs' gestation periods are short, only eight weeks. They suddenly become enormous and the next thing you know, they are giving birth, preferably in front of a good fire, on your hearthrug. After going through twelve months with Kitty and nine with Elly, eight weeks seemed an almost indecently short amount of time.

She may have been heavy but Sam was in great condition I thought, as I threw the first of many stones for her. Her coat was gleaming, a pale, silky blonde and, despite her bulk, she leapt through the thick growth of fionnàn and furze on the hill with her usual mad abandon. Being the mother of thirty-eight pups to date has done nothing to slow her down.

We walked a good way up the mountain, further than I had intended, with the sun warming our backs, then sat on a large, solitary boulder looking out at the greening of the glen. The day was deceptive as March often can be. When the sun suddenly disappeared behind a cloud, a keen, cutting wind sweeping down from the Kenmare Road soon had me up and walking again.

I headed for an ancient patch of woodland, one of four that grace Lickeen, dense areas of oak, birch and gnarled holly. It wasn't far out of my way, and I wasn't ready to go home yet. This part of the land was covered with low, moss-encrusted stone walls that ran everywhere at different angles.

All over Lickeen there were fields that had been shaped by unknown hands, walls that seemingly went nowhere although once upon a time, they must have. You didn't lug all those stones around just for the heck of it. So much was lost now behind layers of moss and in the mists of time. It could never be recalled. But we could create something here again, something different that combined as much of the old as we could save with the new. It was going to take time and lots more money to reclaim this rough land, to come any- where near my dream of a working and productive farm. Somehow I was just going to have to find more of those precious commodities.

I stopped to examine a huge, saucer-shaped fungus which was growing out of the stump of a dead oak tree. It was the size of a dustbin lid and its underside was brightly coloured. I was trying not to think about what I would do if there was a letter in my mail-box saying that the publishers didn't want my book. For once I had no contingency plan.

My doughty old Renault, survivor of much abuse, had indeed got a faulty starter motor. I had stopped at the garage in the village and they'd confirmed my diagnosis. Sometimes it started, sometimes it didn't. When it didn't, you had to jump out, lift the bonnet and severely beat the offending part (when you were able to find it) with a hammer. At first, the light upholstery variety had sufficed, but, as time went on, the problem was getting worse and the hammers heavier. The other day it had taken a sledge-hammer and a crowbar to get any action. The starter motor was probably going to collapse completely under this onslaught and a new one would be expensive. If it gave up on me before I could get the money together I would be stranded up here on this mountain, a quarter of a mile off the road, five miles from any shops and with no way to get essential supplies to the house. I offered up a silent prayer as I headed back down the mountain. Just for today, I asked, let me not get any more bills.

But I must have been facing the wrong direction or

perhaps my prayers had been egocentric for too long. In any event, today they fell on deaf ears. There were several letters. I flicked through them dispiritedly. A solicitation from the Sisters of Mercy, complete with self-addressed envelope, a bill of three-hundred odd pounds for my car insurance and a demand for road tax. Since my cash-flow from features had slowed down again as I'd been working on The Book, there was little chance of any compensating cheques.

I felt weary and heavy-hearted as I headed back up the boreen towards the house. What on earth had made me think I would ever be able to earn a living here in the first place? It had been bad enough trying to keep body and soul together at Barley Lake Cottage. But Lickeen and the animals had stretched me over and beyond my limits. I was going to be up here on this bloody mountain, I thought, kicking savagely at a stone, with no car and no phone either (that bill was due again soon) – a journalist in the middle of nowhere with no means of communication and, let's face it, no career.

I remembered what the cattle dealer had suggested the day I had met him in Glengarriff. Elly would fetch a good price. It would be enough to cover these latest pressing demands and get me off the hook for a while longer. Sadly, I decided to call him that evening.

The house was silent, sombre-looking, when I got back, despite the brightness of the day, as if the ancient, weather-beaten stone was catching and reflecting my own dark mood.

I dragged myself back to the typewriter but the results didn't seem worth the effort – dull, stilted prose. After a couple of hours I gave up in disgust and decided to go downstairs and light the fire early. I would huddle miserably before its warm glow and try to work up the courage to make that call. During the months I had owned Elly, I had come to admire her tough and uncompromising spirit. We had been through a lot together.

There were plenty of other cows but there would be no replacing Electra.

CHAPTER TWENTY-ONE

BEGINNINGS

Knowing Sam would be having her pups soon, I had taken the door off the cupboard under the stairs and made her a bed there. It was a perfect spot – warm, dark and secure. Being a girl who loves her comfort, it was never difficult to persuade Sam to have her pups anywhere you wanted her to, providing you offered her blankets and an old duvet. Otherwise she would pick the most comfortable spot in the house and have them there.

Sam was under the stairs now, stretched out and panting heavily. I set about feeding the other animals and offered Sam her usual dinner but she refused to be enticed out of her snug den. In my experience this meant only one thing. Sam would have her pups that night. Feeling entirely mercenary about it, I prayed for a large, healthy litter. Well-bred dogs that they are, Sam and Tom's offspring fetch good money. It would be a couple of months before they were ready to be sold, but it was money I could count on.

I made myself something to eat and tidied up, keeping a close eye on Sam. She was still breathing heavily but she seemed no further along with things, so I built up the fire and settled down in front of it with a book, ready for a long night. This would be her fourth litter and, old hand that she was, Sam just lay there and got on with it – eventually. She always took a long time giving birth, sometimes as much as six hours. My role was to offer encouragement and to make sure that none of the tiny, newborn creatures got inadvertently squashed in all the excitement.

While I was waiting for something to happen I decided I may as well get the dreaded phone call over with. It wasn't going to become any easier. My cattle dealer friend said he would collect Elly in a couple of days. Prices were good at the moment, he assured me. She would fetch top money at the weekend cattle market.

I felt wretched, as if I were betraying a friend. But that was ridiculous. I was supposed to be a farmer and farmers bought and sold animals all the time. I forced my attention back to Sam just as she produced her first pup. It was already midnight and this seemed exceptionally slow going, even for her. I looked at the tiny damp body from a respectful distance and thought it seemed exceptionally big.

By two o'clock, she had had a second even bigger pup but she had slowed down considerably. At this rate, it would take all night to produce her customary six or seven.

By four o'clock, a third pup had joined the others, big and apparently very healthy too. My eyes were closing but I knew I had to keep awake somehow. There was always a chance that a dead pup might not be expelled and the vet would have to be called. If everything was all right, after the last pup was born Sam would settle down to nurse the others and would usually go to sleep. I always waited around until things were at this point, just to be sure.

By nine o'clock that morning I had become reasonably convinced that there would be no more. The three fat, squirming pups were sucking contentedly and there was much lip-smacking and snuffling coming from under the stairs. Sam was sleeping the sleep of the just.

It was the smallest litter we had ever had, and just at a time when I had been counting on the money. At least we hadn't had any deaths this time though, and the pups were beautiful. I stroked one of the tiny warm bodies and felt that rush of wonder that any birth, animal or human, always provokes in me.

I was too tired to go up to bed, so I stretched out on the hearthrug next to the proud father who had slept through the whole thing. Tom lay close to me and soon his warmth and the steady rhythm of his breathing had lulled me to sleep.

I stayed in that position for hours. When I awoke, the fire was nearly out and the first, grey streaks of nightfall were tinting the sky. I would have to get the milking done and feed the animals before it got much darker. I couldn't believe I had slept so long. I felt confused and disorientated.

Heaving myself off the floor with a reluctant groan I thought I'd see if Sam could be persuaded outside for a few moments before I started on the livestock. She was always a fantastic and protective mother and I knew that to leave her babies for even a few minutes was absolute torture for her. As soon as I managed to tempt her outside she would be whining anxiously to come back in again.

I put on my overcoat and wellies and Tom joined me at the front door. I whistled for Sam, wondering if she was going to take any notice of me. To my surprise, she bounded outside the moment I opened the front door and had a long, well-deserved drink of water. Perhaps she was feeling so relaxed because she had only had three pups this time? She was showing no signs of wanting to go back in.

I decided to make the most of her relaxed attitude and take both dogs with me down to the post-box. I hadn't checked the mail yet today. Sam went down the boreen happily enough. She even chased a few stones that I threw for her. But you could tell that her heart wasn't really in it.

By the time we had got down to the gate, Sam was whining worriedly. I stuffed the couple of envelopes that were inside the postbox into my overcoat pocket and turned for home, Sam racing urgently ahead of me.

The moment I opened the front door, Sam shot inside yelping and disappeared under the stairs. The pups tumbled all over her whimpering ecstatically. Their voices sounded much stronger than they had this morning.

I plugged in the electric kettle to heat water for the animals' concentrated feed. As I was waiting for it to boil, I glanced cynically at the day's crop of post which I had tossed onto the table. The envelopes were damp and crumpled from having been stuffed in my pocket. At first glance they didn't look too bad – a letter from an old and dear friend in Los Angeles, a postcard from someone who was having a wonderful time in India – and a white envelope with an English postmark. At least it couldn't be someone I owed money to.

I measured horse-feed and dairy-nuts into respective buckets and poured the hot water over them. The heady scent of bran, oats and molasses filled the kitchen. While I waited for the mixture to soften, I turned the white envelope over curiously. Stamped in the top right-hand corner was the logo of the publishers I had been waiting to hear from.

I let the envelope drop from my fingers as if it had suddenly started glowing, unable right now to cope with what it might contain. Perhaps I would never be able to cope with what it contained, come to that. It would be naïve and ridiculous of me to expect the cavalry to arrive just when I needed them most. How I would deal with the rejection slip was something I didn't want to think of yet.

I went to the barn with the feed. It was a beautiful night. The sky was a dusky, medieval blue, lightened by the thin sliver of a new moon which appeared to rest delicately on top of the Caha mountains. There were no stars out yet but the sky was clear so there would be. I sat outside the barn for a while, ignoring the animals indignantly calling to me from within.

The cliff which rose steeply above the house looked

somehow alien in the moon's cold clear light. Its sharp peak
cast long strange shadows. I shivered slightly and pulled my
coat around me more tightly.

I fed the animals slowly, filled the hay nets slowly
anything to put off facing that envelope a bit longer. Then
settled down to milk Elly. I took my time, feeling curiously
detached as I performed these familiar yet precious tasks, a
though I were watching myself on screen, moving around
lightly, confidently, in control. The truth was, I'd realized
that this dream could slip away from me, and I was not in
control at all.

I was aware only of the sound of milk gushing into the
pail as I mechanically milked Elly, her steady and efficient
munching as she chomped methodically on the hay, and of
the warm sweet smell of Kitty, standing quietly behind me
watching us. These were the things that mattered to me now
– animals that were warm and well fed near me, a house
loved, a landscape that was so beautiful that at times I felt as
though it might break my heart. Eternal, precious things that
I had wanted for a very long time.

Whatever happened, even if in the end I couldn't make a
go of it, I knew I would be richer for the Lickeen experience
I had learned so much in such a short time. And I had
acquired patience, to a degree I had never possessed before.

I finished milking Elly and checked all her teats to be
sure there was no milk left. Paddy had finished her concen-
trates and was waiting for what she still considered to be the
main course – milk. I tipped half of what I'd just got from
Elly into her bucket. She stuck her head into it greedily.

Inside the house everything would be warm and welcoming. I would pour myself a large whiskey, stoke up the fire
and face the envelope.

I went through a few more elaborate and unnecessary
rituals indoors, winding clocks, moving some papers around
then finally settled myself in the fireside chair and opened the
letter as if it might explode. Tom came over to me and rested

his large head gently on my knee. I rubbed his ears for a moment then took a deep breath and forced myself to read the neatly typed text. I had to go over it three times before I finally realized that the news was good – great, even. My book had been accepted.

I don't know how long I sat there, holding the letter, staring unseeing in front of me. I was at a loss as to how to adjust to such stunning good news.

When I had recovered my senses slightly, I picked up the phone to call Amber. I had promised to let her know as soon as I heard anything.

There was a long way to go yet. I had only written five chapters. Now I would have to put my money where my mouth was and finish it. And when I had done that, would anybody want to buy it? Meanwhile, there would be bills to pay, land to be reclaimed, and animals to care for. Logically, Lickeen was still a huge gamble and there could be many problems ahead of me. But then, our decision to buy Lickeen had been based on intuition, not logic. I'd abandoned that commodity a long time ago.

But there was one major and significant factor here that would change everything. There was hope, a way of making the disparate strands that were my life come together at last. Everything I had worked for during those years at the *National Enquirer*, the other jobs I'd had over the years that I hadn't particularly enjoyed were contained here. I had put down roots that already went deep.

Something else dawned on me. Now I wouldn't have to sell Elly. I could call the cattle dealer tomorrow and tell him not to come. I couldn't get over the feeling of relief.

As I waited impatiently for Amber to answer the phone, I realized that this was only the beginning.

Lickeen – this frustrating, infuriating, exhilarating, irresistible place – had not finished with me yet. But that's another story for another day . . .